Finite Element Analysis - Applications in Mechanical Engineering

Finite Element Analysis - Applications in Mechanical Engineering

Editor

Albert Brose

scitus
academics

Finite Element Analysis - Applications in Mechanical Engineering
Edited by **Albert Brose**

ISBN: 978-1-68117-309-2
Library of Congress Control Number: 2016937618

SCITUS Academics LLC,
www.scitusacademics.com
Box No. 4766, 616 Corporate Way,
Suite 2, Valley Cottage,
NY 10989

Notice

Printed in United States of America on Acid Free Paper

Preface

Finite element analysis (FEA) has become a key technology in the modeling and simulation of various engineering systems in recent decades. FEA is a computational tool for engineering analysis. The use of mesh-generating techniques to divide a complex problem into small elements and the use of FEM-coded software includes. The complex problem in the application of FEA is usually a physical system with the underlying physics such as the Euler-Bernoulli beam equation, the heat equation or the Navier-Stokes equations expressed in partial differential equations or integral equations, while the small divided elements of the complex problem represent different areas of the physical system. FEA is a computerized method to predict the reaction of a product to real world forces, vibration, heat, fluid flow and other physical effects. The analysis of finite elements indicates whether a product breaks, wears or operates as intended. Analysis is called but is used to predict what happens when the product is used in the product development process. FEA divides a real object into several finite elements, such as small cubes. Mathematical equations help to predict the behavior of each element. A computer then adds all the various behaviors to predict the behavior of the real object. The use of the finite element method to solve engineering problems involves the analysis of linear elastic stress, thermal analysis and limitations and errors of modeling. Commercial computer codes are used in applications. This book covers on recent FEA applications in main areas of mechanical engineering.

Table of Contents

1

CHAPTER

A NEW METHOD COMBINING FINITE ELEMENT ANALYSIS AND DIGITAL IMAGE CORRELATION TO ASSESS MACROSCOPIC MECHANICAL PROPERTIES OF DENTIN

Wenlong Wang [1], Nicolas Roubier [1], Guillaume Puel [1], Jean-Marc Allain [2], Ingrid C. Infante [3], Jean-Pierre. Attal [4] and Elsa. Vennat [1,*]

[1] MSSMat, UMR 8579 CNRS-Ecole Centrale Paris, Grande Voie des Vignes, 92295 Chatenay-Malabry Cedex, France
[2] LMS, UMR 7649 CNRS-Ecole Polytechnique, 91128 Palaiseau Cedex, France
[3] SPMS, UMR 8580 CNRS-Ecole Centrale Paris, Grande Voie des Vignes, 92295 Chatenay-Malabry Cedex, France
[4] URB2i, EA4462 Université Paris Descartes, 1 rue M. Arnoux, 92120 Montrouge, France

ABSTRACT

A literature review points out a large discrepancy in the results of the mechanical tests on dentin that can be explained by stress and strain assessment during the tests. Errors in these assessments during mechanical tests can lead to inaccurate estimation of the mechanical properties of the tested material. On top of that, using the beam theory to analyze the bending test for thick specimens will increase these experimental errors. After summarizing the results of mechanical tests on dentin in the literature, we focus on bending tests and compare the stress assessment obtained by finite element analysis (FEA) and by beam theory application. We show that the difference between the two methods can be quite large in some cases, leading us to prefer the use of FEA to assess stresses. We then propose a new method based on coupling finite element analysis and digital image correlation (DIC) to more accurately evaluate stress distributions, strain distributions and elastic modulus in the case of a three-point bending test. To illustrate and prove the feasibility of the method, it is applied on a dentinal sample so that mean elastic modulus and maximum tensile stress are obtained (11.9 GPa and 143.9 MPa). Note that the main purpose of this study is to focus on the method itself, and not to provide new mechanical values for dentin. When used in standard mechanical testing of dentin, this kind of method should help to narrow the range of obtained mechanical properties values.

Keywords: dentin; three-point bending test; digital image correlation; finite element analysis; mechanical properties; stress and strain assessment

1. INTRODUCTION

Dentin is the main mineralized biological tissue of the tooth. It is located between the enamel and the pulp cavity and shows a hierarchical and complex structure at different scales. At the nanoscale, it consists of a carbonated nanocrystalline apatite mineral phase (approximately 50% by volume), a grid work of type I collagen fibrils (approximately 30% by volume) and fluids (approximately 20% by volume) [1]. At the microscale, dentin can be seen as a continuous fiber-reinforced composite with peritubular cuffs as reinforcement and a matrix of intertubular dentin. At the macroscale, it can be seen as a bulk material with effective properties resulting from its complex structure. The knowledge of these properties is crucial to predict the tooth's response to applied loads [2]. Dentin mechanical properties assessment is also important in order to improve its restoration and to realize more biomimetic restorative materials.

The specific microstructure of dentin described above obviously influences the dentinal mechanical properties. Considering dentin as a composite made of tubules, peritubular and intertubular dentin, as shown in Figure 1, dentin can be approximately modeled as a transverse isotropic material [2,3].

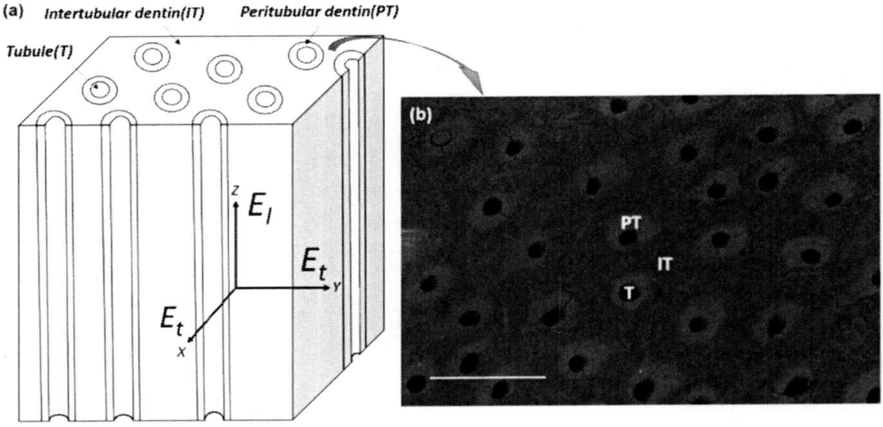

Figure 1. (**a**) Dentinal structure: the directions of longitudinal modulus (E_l) and transverse modulus (E_t) are shown in the scheme. (**b**) Scanning electron microscopy image from the top view of dentin, with the identification of regions of peritubular dentin (PT), intertubular dentin (IT), and a tubule (T). Scale bar corresponds to 10 μm.

Thus, the relation between stress **σ** and strain ε in linear elasticity could be described by the classical relationship [4]:

$$\begin{bmatrix} \varepsilon_{xx} \\ \varepsilon_{yy} \\ \varepsilon_{zz} \\ 2\varepsilon_{yz} \\ 2\varepsilon_{xz} \\ 2\varepsilon_{xy} \end{bmatrix} = \begin{bmatrix} \frac{1}{E_t} & -\frac{\nu_{tt}}{E_t} & -\frac{\nu_{tl}}{E_t} & 0 & 0 & 0 \\ -\frac{\nu_{tt}}{E_t} & \frac{1}{E_t} & -\frac{\nu_{tl}}{E_t} & 0 & 0 & 0 \\ -\frac{\nu_{tl}}{E_t} & -\frac{\nu_{tl}}{E_t} & \frac{1}{E_l} & 0 & 0 & 0 \\ 0 & 0 & 0 & \frac{1}{G_{tl}} & 0 & 0 \\ 0 & 0 & 0 & 0 & \frac{1}{G_{tl}} & 0 \\ 0 & 0 & 0 & 0 & 0 & \frac{2(1+\nu_{tt})}{E_t} \end{bmatrix} \begin{bmatrix} \sigma_{xx} \\ \sigma_{yy} \\ \sigma_{zz} \\ \sigma_{yz} \\ \sigma_{xz} \\ \sigma_{xy} \end{bmatrix}$$

where E_t, E_l are the transverse and longitudinal modulus respectively, ν_{tl}, ν_{tt} are the associated Poisson's ratios, and G_{tl} is the shear modulus in longitudinal direction.

These transverse and longitudinal moduli but also the maximal tensile and compressive stresses have been widely investigated experimentally since the middle of the last century [2], using either compression or traction experiments. Through tensile tests, the elastic modulus (E_t) have been estimated by Sano et al. [5] to be 13–15 GPa. When using an optical strain gauge to determine the strain [6], the elastic modulus was reported to be 19.3 GPa but with a larger deviation (28%). Through compressive tests, the elastic modulus of human dentin was determined by Peyton et al. [7] to be 11.6 GPa. Strain gauges were attached to steel rods that were then used to apply the load to the specimens. A compressive modulus value of around 14 GPa can be found in the literature [8,9]. These lower moduli values obtained in compression may arise from the non-parallel alignment of load and possible relaxation effects. Craig and Peyton [10] reported a modulus of 18.5 GPa modified by load–unload cycles with the strain gauge directly onto the sample to remove the effect of the compression setup rigidity. Recently, Zaytsev et al. [11] reported elastic modulus values within a range of 3 to 13 GPa as a function of the geometry of the sample. However, they did not explicitly indicate the method used for the strain measurement, which made it difficult to place their results within the framework of others. Finally, diametral compressive and classical compressive tests were performed by Palamara et al. [8] using digital image correlation (DIC) to determine elastic modulus and the values were 6.5 ± 2 GPa and 10.7 ± 2.4 GPa respectively. In all these tests, from the distinct strain rates we deduce that the results are likely to be influenced by dentin viscoelasticity: Jantarat et al. [12] have already reported a linear dependence of elastic modulus as a function of strain rate.

Bending tests seem to be more rarely used, probably due to the fact that a special setup is needed to carry out such tests. Nevertheless, three-point [13] or four-point [14] bending tests can be used to assess dentin mechanical properties. Rees et al. [13] determined a static elastic modulus (of 8.6 GPa with a standard deviation of 0.86 GPa) using a three-point bending test and a dynamic modulus ranging from 14.3 to 15.9 GPa in the frequency range of 0.1 to 10 Hz. Ryou et al. [14] obtained an elastic modulus (13–17 GPa) depending on the location in dentin with four-point bending tests.

Some other techniques were also used to determine dentin mechanical properties. Transverse isotropic elastic constants along orthogonal directions to the tubules (E_t) were indeed studied by Lees and Rollins [4] using critical reflectance ultrasound spectroscopy (RUS) measurements. A modulus of 36 GPa in the tubule direction (E_l) and 29 GPa in the perpendicular axis (E_t) was reported. Differing values (from 23 to 25 GPa) were reported by Kinney et al. [3] using resonant ultrasound spectroscopy measurements, but without assumption on material symmetry. More recently, anisotropic properties depending on the tubule orientation were investigated through micro-pillar compression tests [15–17], the moduli were reported from 3.5 to 16 GPa as a function of the tubule orientation. In this study, ten tubules per pillar can be observed, which might be slightly lower than what should be found in a Representative Volume Element (RVE) [18].

Table 1 summarizes the different values of dentin elastic modulus and ultimate strength found in the literature. We note that a wide range of values can be found and a great number of methods for stress and strain assessments have been used, indicated if available.

Table 1. Literature review of the mechanical properties of human dentin (elastic modulus units: GPa, stress unit: MPa).

Methods	E_l	E_t	$\sigma_{bending}$	$\sigma_{compression}$	$\sigma_{tension}$	ε measure	σ measure
RUS [3]	25.0	23.3					
RUS [4]	36.0	29.0					
Micro-pillar compression [16]	13.0	3.5		60–160		$\varepsilon = \frac{d}{L}$	$\sigma = \frac{F}{S}$
Micro-pillar compression [17]	16.2	13.2		186–210		$\varepsilon = \frac{d}{L}$	$\sigma = \frac{F}{S}$
Diametrical compression [8]	6.5 ± 2.0	6.5 ± 2.0			50.9–58.7	DIC	$\sigma_{compression} = \frac{2F}{\pi D t}$
Compression [7,8,10]	10.7 ± 2.4	11.9 ± 3.0		294–333 [8] 230–300 [7,10]		DIC [8] Strain gauge [7,10]	$\sigma = \frac{F}{S}$
Three-point bending [13]		8.7 ± 0.86				Crosshead (deflection assessment)	Beam equation
Four-point bending [19]			145–326			Crosshead (deflection assessment)	Beam equation
Tension [1,5,6]		6.0–19.3			30–130	Strain gauge	$\sigma = \frac{F}{S}$

d is the crosshead displacement; L is the length of sample in compression direction; F is the load from the compression device; S is the cross section area of the specimen; D is the diameter of cylindrical sample; t is the thickness of cylindrical sample.

In order to narrow the range of dentin mechanical properties values, a new method to assess stress and strain in the case of a three-point bending test is proposed. A two-step procedure is thus chosen:

First, for typical, published, bending tests, we propose to re-evaluate the stresses by the finite element method instead of using the Bernoulli–Euler beam theory as it is classically done. The differences in terms of maximum stress will be illustrated through the analysis of two articles [13,19].

Secondly, a three-point bending test combined with Finite Element Analysis (FEA) and Digital Image Correlation (DIC) is proposed, taking advantage of the complementarity of these two methods.

The DIC technique is increasingly used to investigate the mechanical behavior of human tissues [20] in that it enables the obtainment of the whole strain field on a surface. However, this technique is not commonly used in dental material testing: to our best knowledge, only two studies on dental materials used DIC [8,20]. At the same time, FEA allows the most accurate evaluation of stresses in complex configurations but has not been used to assess stresses during dentin mechanical testing to the authors' knowledge. Surprisingly, there have been few studies using this combination of two methods (FEA and DIC) to determine the elastic modulus of dentin.

2. EXPERIMENTAL SECTION

2.1. Recalculation of the Stresses in the Bending Test Configurations Found in the Literature

Two articles were chosen because they provide sample geometries and some results from three-point bending tests and four-point bending tests respectively [13,19]. The values from the published articles were recalculated using the classical beam theory (Sections 2.1.1 and 2.1.2). Then, using the experimental conditions mentioned in the two selected articles, the stresses were recalculated by the finite element method (Section 2.1.3).

2.1.1. Three-Point Bending Test [13]

Rees et al. [13] performed static three-point bending tests to measure dentin elastic modulus (Figure 2).

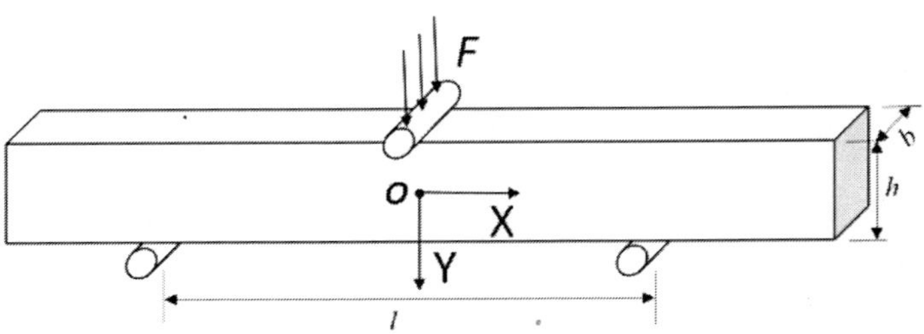

Figure 2. Scheme of three-point bending test.

To assess the stresses in the beam, the Bernoulli–Euler beam model can be used (at x = 0, which is the cross-section where the stresses are the highest) with:

$$\sigma_{xx} = \frac{3Fl}{bh^3}y$$

where σ_{xx} is the normal stress in direction x, F is the load, l is the supporting span, b is the beam width, h is the beam thickness, and y is the vertical coordinate.

In the Rees et al. experiment [13], beams were 2 mm wide, 2 mm in depth and presented an average length of 13 mm. The distance for the supporting span was 11 mm. Thus, the ratio (l/h) was 5.5. Then, the modulus was calculated from Bernoulli–Euler beam equation:

$$E = \frac{Fl^3}{4bdh^3}$$

where F is the load, and d is the deflection at the given load.

2.1.2. Four-Point Bending Test [19]

Eltit et al. [19] performed four-point bending tests on dentinal samples. They studied the maximum flexural stress of coronal and root dentin. The tested beam dimension was 0.4 mm × 1.2 mm × 7 mm (thickness × width × length). The supporting span of the bending device was 5 mm, and the loading span was 1.5 mm, as is shown in Figure 3.

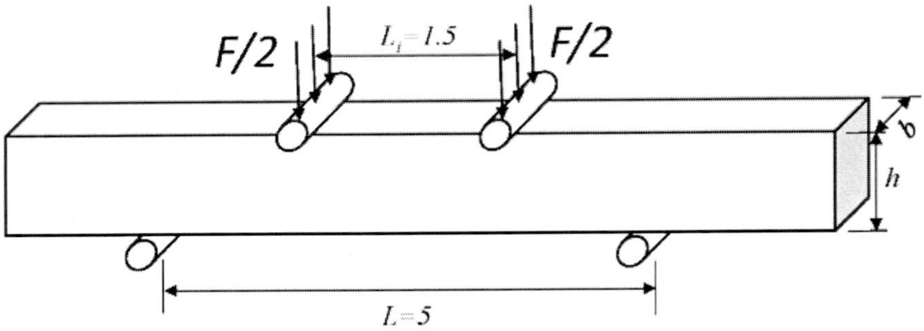

Figure 3. Scheme of four-point bending test.

According to the Bernoulli–Euler beam equation, the maximum flexural stress can be expressed as:

$$\sigma_f = \frac{3F(L - L_i)}{2bh^2}$$

where σ_f is the maximum flexural stress, F is the load, L is the supporting span, L_i, is the loading span, b is the beam width, and h is the thickness of the beam.

2.1.3. Recalculations by Finite Element Analysis (FEA)

The simulations were performed with the commercial software COMSOL Multiphysics (Comsol Co., Stockholm, Sweden). In the FEA, we use the static equilibrium equations under small deformation assumption with an isotropic relation:

$$\boldsymbol{f_b} + \nabla \cdot \boldsymbol{\sigma} = 0$$

$$\varepsilon = \frac{1}{2}[\nabla \boldsymbol{u} + \nabla \boldsymbol{u}^T]$$

$$\varepsilon = \frac{1+\nu}{E}\boldsymbol{\sigma} - \frac{\nu}{E}tr(\boldsymbol{\sigma})\mathbf{I}$$

where $\mathbf{f}_b$ is the body force, $\boldsymbol{\sigma}$ is the stress tensor, ε is the strain tensor, u is the displacement, ν is the Poisson ratio, E is elastic modulus, and **I** is the identity matrix.

For recalculating the stress in three-point bending test, about 204,000 tetrahedral elements were used in the mesh. The boundary conditions are the following:

- Fixed constraints (u = 0) were applied to the two support edges.
- Boundary load of total force (50 N) was applied to the middle area (0.6 mm × 2 mm) between the two supporting edges.

For recalculating the stress in four-point bending test, about 240,000 tetrahedral elements were used in the mesh. The boundary conditions are the following:

- Fixed constraints (u = 0) were applied to the two support edges.
- Edge load of total force (5 or 12 N) was applied to the loading part (consisting of two loading edges here).

The edge load was chosen in order to reproduce data from the Eltit et al. [19] experiment. Within the load-deflection curves from the four-point bending tests of Eltit et al. [19], it can be inferred that the maximum bending force used for crown dentin beams was about 5 N, and for root dentin beams, the bending force was around 10–15 N (here, 12 N was chosen as an example).

2.2. Proposed Experimental Protocol

2.2.1. Sample Preparation

A dentin specimen was obtained from a sound human third molar. The specimen was polished progressively by 80, 500, 800, 1200,

1400, 2000, 2400, 4000 grit SiC polishing papers under water irrigation, and finally turned to cloth polishing with 1 μm diamond suspension liquid for both sides. Then, the dentin slice was cut into one beam with a line cutting saw (Isomet low speed saw, BUEHLER Co., Lake Bluff, IL, USA) (Figure 4). After cutting, the new surfaces were then polished with the same previous polishing steps. Finally, the beam dimensions were controlled as 8 mm × 1.38 mm × 1.15 mm. Before each step, the sample was washed in an ultrasonic cleaner for 3–5 min. Then, the specimen was stored in the distilled water below 4 °C until the test.

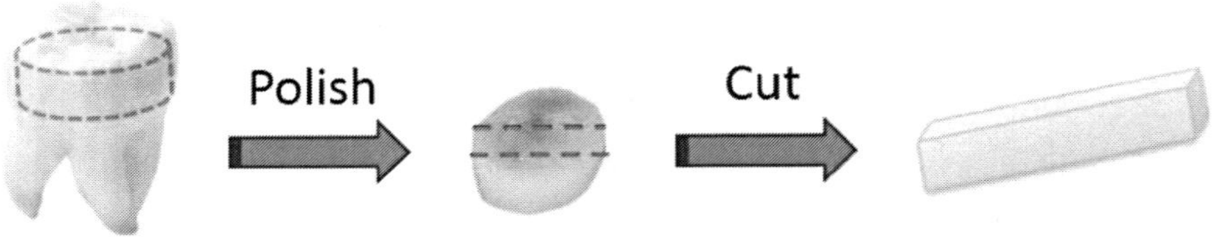

Figure 4. Scheme of sample preparation.

2.2.2. Three-Point Bending Setup

Three-point bending tests were performed using a bending test device equipped with a 1 kN load cell (MICROTEST, Deban UK, Suffolk, UK) with a supporting span of 5 mm. Motor speed was selected as 0.2 mm/min. All the setup was mounted under an optical microscope (InfiniteFocus, Alicona Co., Raaba, Austria) in order to get the strain distribution of the beam side surface located below the force-indenter as shown in Figure 5. The direction of the load is applied perpendicular to the tubular orientation.

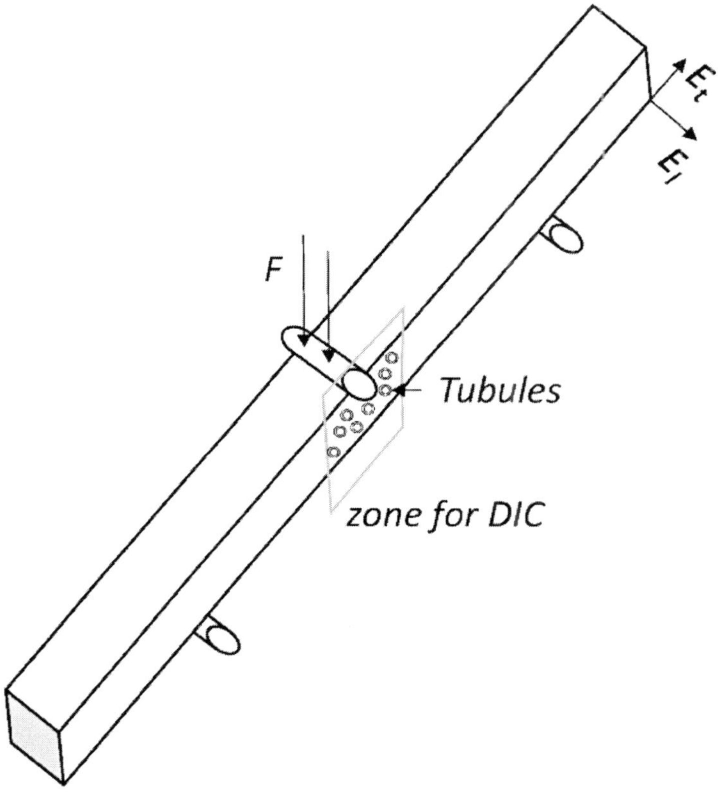

Figure 5. Three-point bending test under the optical microscope.

2.2.3. DIC

The software correli_q4 [21] was used to analyze the strain distribution comparing the reference and deformed images. The deformed image was taken under the load force of 60 N. The element size for DIC was selected as 64 pix.

2.2.4. Determination of the Elastic Modulus and σ_{max}

The mechanical setup is placed under an optical microscope in order to get images of a selected area of the sample (Figure 5) at different loads. Strain estimates are obtained by DIC and the stresses are estimated by FEA. The edge load can be known from the three-point bending test device. With the maximum total force, one can obtain the maximum tensile stress by FEA. During the bending test process, the strain is evaluated by DIC. Using stress and strain estimates in direction x as shown in Figure 1, when the terms σ_{yy} and σ_{zz} are much lower than σ_{xx}, the elastic modulus in direction x (E_t) can be derived by Hooke's law:

$$E_t = \frac{\sigma_{xx}}{\varepsilon_{xx}}$$

In our case, the maximum total force found in the experiment (F_{max} = 65 N) was used to determine the maximum bending stress. Tensile test and bending test are known to give slightly different results even when the material fails in tension in both tests, which is why the bending test results in term of strength are called flexural or bending strength and not tensile strength. In this article, we will estimate the maximum tensile stress in the experiment and call it "flexural strength", as is usually done.

A total force of 60 N (F_x) was used to determine the stress in order to obtain the elastic modulus, as shown in Figure 6.

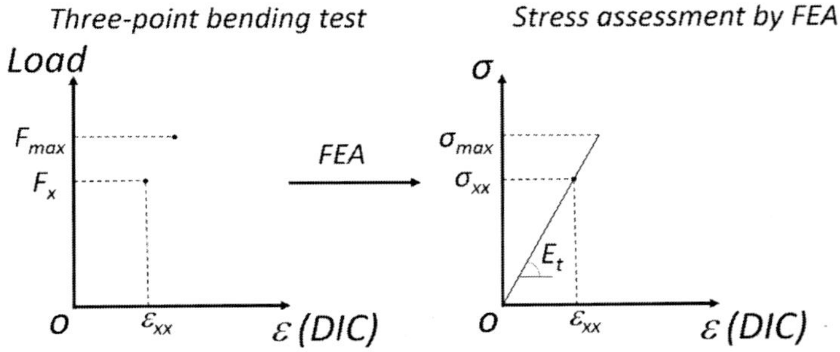

Figure 6. Method of the determination of elastic modulus and of the maximum stress values σ_{max}.

3. RESULTS AND DISCUSSION

3.1. Recalculation of the Stresses in Bending Test Configurations Found in the Two Selected Articles

3.1.1. Three-Point Bending Test [13]

With the beam dimensions in three-point bending test of Rees et al. [13], using the Bernoulli–Euler equation, the maximum tensile stress (often called maximum bending or flexural stress) with assuming boundary force (F = 50 N) is calculated as 103.1 MPa, whereas, using FEA, the maximum tensile stress is 58.2 MPa. Figure 7 shows the distribution of σ_{xx} in three-point bending test. As expected, there is a neutral plan in the middle of the beam, and the values are antisymmetric relative to the stress direction (compression and tension). The maximum tensile stress is located at the outside surface of the beam.

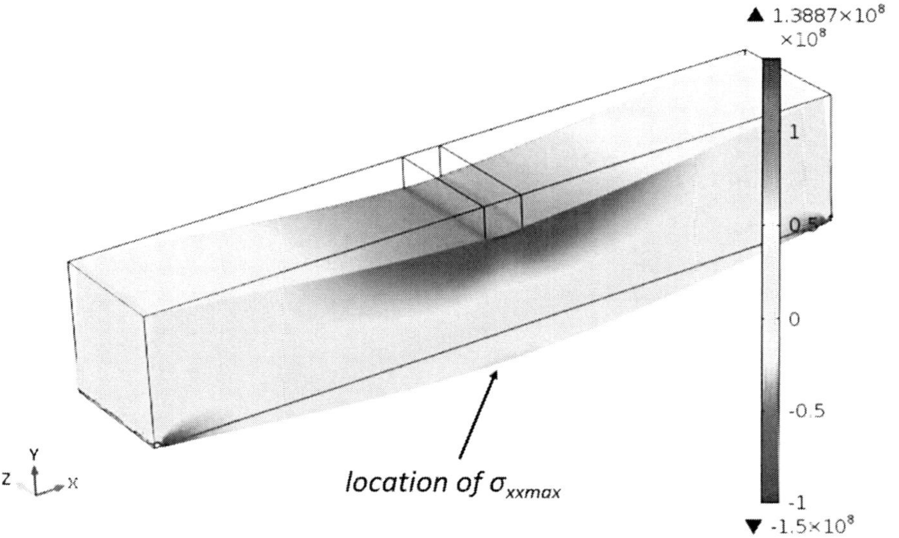

Figure 7. The σ_{xx} distribution in the beam (unit for color scale bar: Pa).

3.1.2. Four-Point Bending Test

According to the Bernoulli–Euler beam equation, the maximum tensile stress for crown and root dentin samples (loads of 5 and 12 N, which are the maximum loads found by Eltit et al. [19]) were 136.7 and 328.1 MPa, which matched the values expressed in [19]. However, by using FEA, the maximum tensile stress for coronal and root dentin test is found as 59.1 and 141.8 MPa, respectively. The σ_{xx} distribution map was shown in Figure 8 with an edge load of 12 N. The maximum bending stress can be found in the outside edge of the beam withstanding tensile stress.

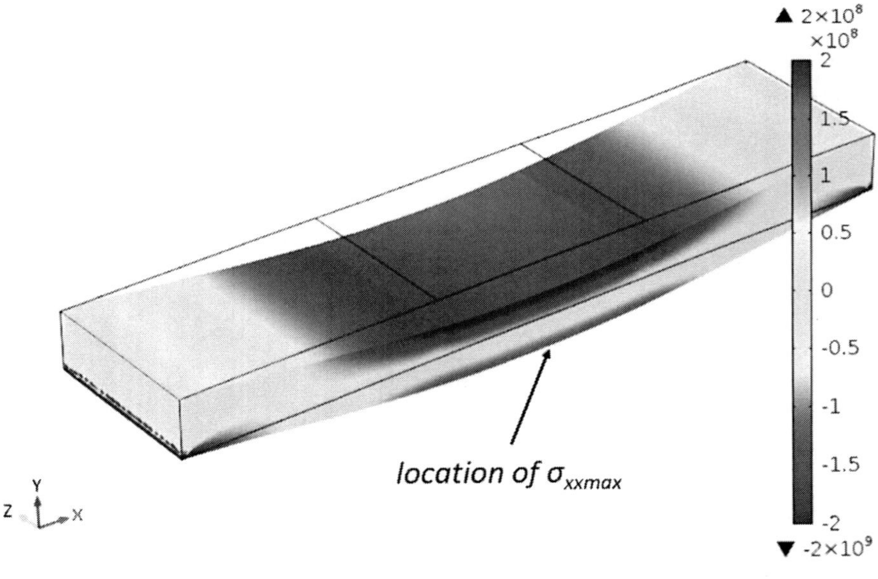

Figure 8. The σ_{xx} distribution map of four-point bending test (unit of color scale bar: Pa).

3.2. Experimental Results

The strain distribution map along direction y obtained by DIC is shown in Figure 9. Note that ε_{xx} shows a linear relationship with respect to the local coordinate in direction y. Five areas (columns 1 to 5) were selected to evaluate ε_{xx} distribution, which is approximately antisymmetric relative to the middle neutral plan as expected. In the same positions, stress distributions were assessed by FEA. Thus, the mean elastic modulus was found as 11.9 GPa.

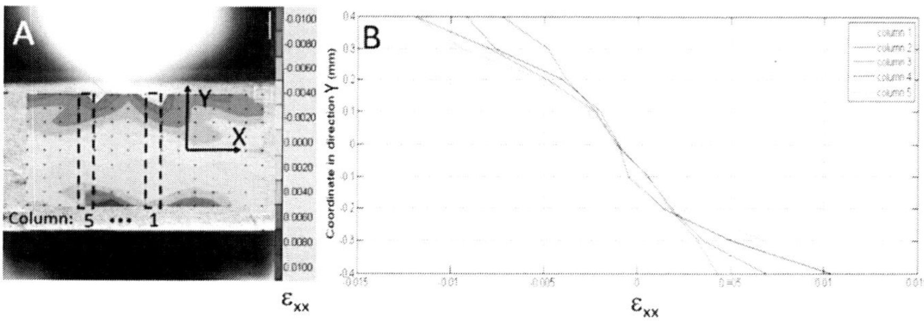

Figure 9. DIC strain distribution for intact dentin beam. (**A**) Three-point bending test and selected areas to acquire ε_{xx} using DIC; (**B**) ε_{xx} distribution along direction y (from DIC).

With the maximum boundary load of 65 N found in the experiment, the stress distribution map is obtained by FEA with beam dimensions corresponding to our dentin sample. Tensile stresses are logically located under the neutral plan. On the location marked in Figure 10, σ_{xx} was found as 143.9 MPa, and at this location σ_{yy} and σ_{zz} were found to be below 0.4 MPa, which were much lower than σ_{xx}, hence the possibility to apply the formula $Et=\sigma xx\varepsilon xx$.

In order to validate our method, the deformations measured by DIC and FEM (using our experimental result and the result from Rees et al. (1994) [13]) were compared (Figure 11). Strain distribution along y coordinates under the indenter (column 2 in Figure 9) measured by DIC was selected as an example.

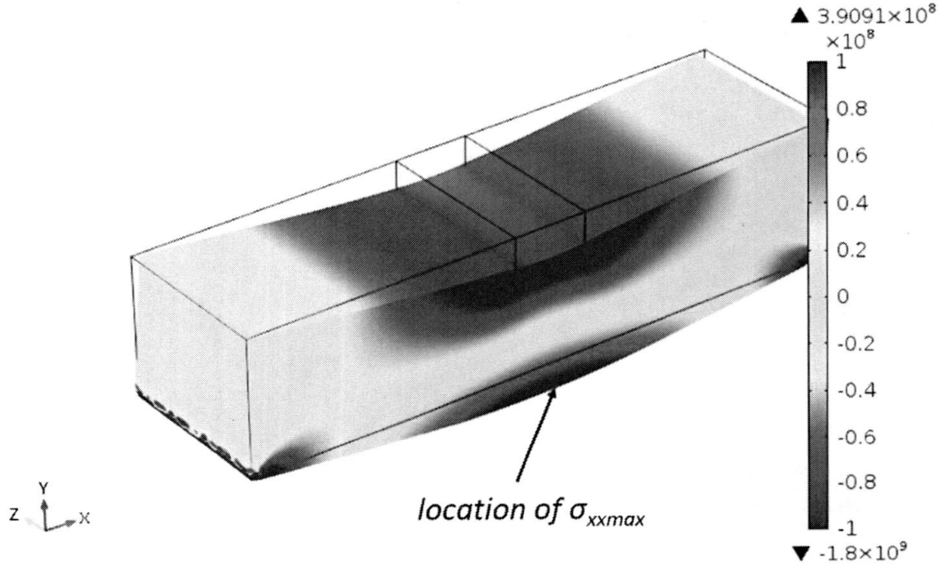

Figure 10. The σ_{xx} distribution estimation (unit of color scale bar: Pa).

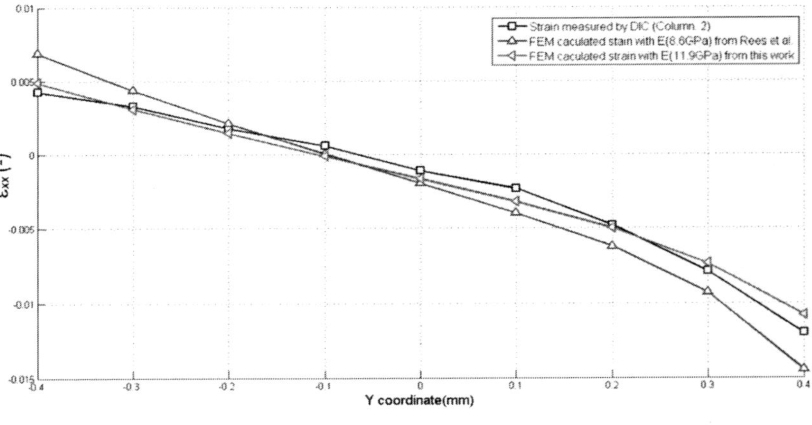

Figure 11. The comparison of ε_{xx} between FEM and DIC along the y coordinate.

3.3. Discussion

Large discrepancies have been found in the literature concerning dentin mechanical properties (and especially its Young's modulus). It is probably caused by stress and strain assessment errors. Indeed Bonfield and Datta [22] compared the value of elastic modulus for compact bone with two kinds of micro-strain measuring techniques: they have shown that with different strain measurements, the value of elastic modulus clearly varied. It is likely to be the same while assessing dentin mechanical properties. The elastic modulus is also very sensitive to the strain rate. As reported by Jantarat et al. [12] and McElhaney [23], great differences of elastic modulus can be observed due to the strain rate variation (from 0.001 to 1500 per second).

From the recalculation of selected articles, it can be inferred that the Bernoulli–Euler beam equation is probably used to get the flexural strength and elastic modulus. Since the Bernoulli–Euler beam theory is based on the hypothesis of a large slenderness ratio, it can lead to an inaccurate assessment of stresses when the slenderness ratio is inadequate. This would also lead to an error of evaluation of the modulus. Compared with the results of four-point bending test of Eltit et al. [19], the maximum bending stress of root dentin obtained from classical beam equation was reported at more than 300 MPa. This is in contradiction with the maximum tensile stress deduced from uniaxial tension test [5,6,8]. However, the maximum bending stress of 55 and 132 MPa (for root and coronal dentin, respectively) deduced using FEA stress assessment seems to better match the results from the literature value, which reported a value of 31 to 130 MPa under uniaxial tension tests [1,5,6]. Thus, it can be inferred that with such sample dimensions, it is inaccurate to use the classical beam theory to assess the stress distribution inside the beam, and that would also lead to an incorrect estimate of the elastic modulus. Nevertheless, elastic modulus determined with beam theory can be used as a comparative tool.

For stress determination, the geometry and positioning of sample are crucial. Some non-parallel alignment of the load can occur due to poorly controlled geometry of the sample in uniaxial compression. In order to well bond the sample with the grips of the machine, an adhesive glue can be used in the tension test [5], at the cost of increasing the relaxation effects. Those effects are not taken into account when using the Bernoulli–Euler beam equation to assess stresses but FEA can also well avoid the influence of those effects.

For strain determination, Sano et al. [5] and Rees et al. [13] used the load-displacement curve from the testing device to get the strain and elastic modulus. Lower values of elastic modulus were reported, which were probably caused by the inaccurate evaluations of strain: the measured displacement includes not only the whole sample displacement but also motions of the loading setup. This leads to the impossibility of comparing the results in different studies done on different machines. In our study, DIC was used to assess the strain on the external surface of the beam for three-point bending. It allows a more global assessment of strain (on a whole surface) with a satisfactory resolution. The accuracy of the DIC technique is determined by the optical resolution of the focusing system used for the image capture. The precision, therefore, mainly depends on the pixel resolution of the camera. The camera in the microscope used in our test has a size of 1280 (horizontal) by 1024 (vertical) pixels, which is higher than in the study of Palamara et al. [8], and thus may lead to a higher resolution on the strain measurement. Moreover, according to Figure 11, we can observe that the values from our finite element model match well with our experimental results.

As discussed above, stress and strain assessment is crucial to get accurate estimates of dentin mechanical properties. For three-point bending test, combined FEA for stress estimate and DIC for strain determination may be a method to explore the dentin elastic modulus and maximum tensile stress. Elastic modulus determined by our method is E_t (because the tensile stress direction is perpendicular to the tubular orientation); our test allowed us to measure E_t as 11.9 GPa, and the maximum stress as 143.9 MPa. The elastic modulus measured here matches the results in the references [5,6,8], in which elastic moduli were reported between 6 and 19 GPa (determined by tension tests). Similarly, our maximum tensile stress corresponds to the magnitude of uniaxial tension test results [1,5,6].

However, for dentin sample, the location in the tooth from where the sample comes may significantly influence the properties of the mechanical test. The density of the mineral phase indeed varies due to the change in tubule density and diameter in the different dentin locations. The statistical structure characteristics have been summarized by Marshall et al. [1]. The percentage of peritubular dentin in crown dentin varies from 60% near the pulp cavity to 3% at the dentin–enamel junction (DEJ), and the percentage tubule area and diameter vary from about 22% and 2.5 μm near the pulp to 1% and 0.8 μm at the DEJ. The intertubular matrix area varies from 12% at the pulp cavity up to 96% near the DEJ. It can be inferred that there is a sharp variation in the density and diameter size of tubules and of peritubular dentin in the different locations from the pulp to the DEJ [24]. Hence, the variations in the properties of the dentin sample may be explained probably by the mineral phase density changes within the tooth. Tests on bone specimen have found that the elastic modulus is proportional related to the mineral density [25]. As local mineral densities are different in the parts of peritubular and intertubular dentin, indentation tests on dentin samples have shown that the elastic modulus strongly depends on the location [17]. Thus, the bending test measured here can be done on different samples taken from different depths to assess the variation of E_t with depth (and also with mineral density).

To model the dependency of the dentin modulus with its components ratio, a classical composite model can be used. To do this, the local moduli are needed. With an indentation system, Kinney et al. [26] have reported significantly different elastic modulus for peritubular (29.8 GPa) and intertubular (18–21 GPa). According to the composite modulus model, elastic modulus in direction x can be described as:

$$E_t = \frac{E_p E_i}{E_p r_i + E_i r_p}$$

where E_p and E_i are the elastic modulus of peritubular and intertubular dentin, respectively, and r_i and r_p are the area ratios of intertubular and peritubular dentin respectively. Without considering the percentage of tubular area, an approximation of the elastic modulus in direction x can be inferred between 18 and 25 GPa. However, the density of the tubules changes a lot within the tooth. With the addition of the tubular ratio in the composite model, the elastic modulus becomes:

$$E_t = \frac{1}{\frac{r_p}{E_p} + \frac{r_i}{E_i} + \frac{r_{tubule}}{E_{tubule}}}$$

where r_{tubule} and E_{tubule} are ratio and modulus for the tubule respectively. Since a very low modulus is expected for the tubules, we can expect a significant decrease of E_t.

Besides, lower values of elastic modulus on dentin have also been found by Marshall et al. [1]. The elastic moduli should be more reasonably represented as a range of values rather than one absolute value, or more accurate values should be obtained to show the relation between the exact location within the tooth (and also mineral density).

It is to be noticed that other factors also influence dentinal mechanical properties, such as storage conditions, tooth age and disease. This is why it is crucial to try to standardize the mechanical tests on dentin in order to be able to compare results of different research groups on different dentin samples (storage, age, disease) and make better evaluations.

4. CONCLUSIONS

Errors in stress and strain assessment during mechanical tests can possibly lead to a misevaluation of the mechanical properties. In this paper, a review of dentinal mechanical properties has been done and large discrepancies were found in the literature.

Two articles of the literature using bending tests were selected and re-examined using FEA to highlight the importance of stress and strain evaluation to get mechanical properties. Then, a new method (using FEA and DIC) is proposed to better evaluate stress and strain distributions, and consequently elastic modulus. To illustrate and prove the feasibility of the method, the elastic modulus and maximum flexural stress were obtained for a dentinal sample, which were respectively 11.9 GPa and 143.9 MPa. The next step of this study is to improve the stress field assessment in the beam by considering dentinal anisotropy in FEA.

Using accurate and robust methods to assess stress and strain is needed to avoid setup-dependent measurements. Keeping this idea as the ultimate objective, an experimental protocol based on a three-point bending experiment under an optical microscope is proposed, which should perhaps trigger discussion on the standardization of dentinal sample mechanical testing. Being able to compare results of different research groups will indeed lead the community to a better understanding of the differences arising from other relevant factors such as species, age, disease, storage condition, etc. (instead of comparing indirectly their equipment).

ACKNOWLEDGMENTS

This work has been benefited from the financial support of the LabeX LaSIPS (ANR-10-LABX-0040-LaSIPS) managed by the French National Research Agency under the "Investissements d'avenir" program (n° ANR-11-IDEX-0003-02).

AUTHOR CONTRIBUTIONS

Wenlong Wang, Elsa Vennat, Guillaume Puel, Nicolas Roubier, Jean-Pierre Attal conceived and designed the experiments; Wenlong Wang performed the experiments; Jean-Pierre Attal provided reagents and materials; Wenlong Wang, Nicolas Roubier, Elsa Vennat, Guillaume Puel, Ingrid C. Infante, Jean-Marc Allain wrote, discussed and revised the manuscript.

CONFLICTS OF INTEREST

The authors declare no conflict of interest.

REFERENCES

1. Marshall, G.W.; Marshall, S.J.; Kinney, J.H.; Balooch, M. The dentin substrate: Structure and properties related to bonding. J. Dent **1997**, 25, 441–458.

2. Kinney, J.H.; Marshall, S.J.; Marshall, G.W. The mechanical properties of human dentin: A critical review and re-evaluation of the dental literature. Crit. Rev. Oral Biol. Med **2003**, 14, 13–29.

3. Kinney, J.H.; Gladden, J.R.; Marshall, G.W.; Marshall, S.J.; So, J.H.; Maynard, J.D. Resonant ultrasound spectroscopy measurements of the elastic constants of human dentin. J. Biomech **2004**, 37, 437–441.

4. Lees, S.; Rollins, F., Jr. Anisotropy in hard dental tissues. J. Biomech **1972**, 5, 557–566.

5. Sano, H.; Ciucchi, B.; Matthews, W.G.; Pashley, D.H. Tensile properties of mineralized and demineralized human and bovine dentin. J. Dent. Res **1994**, 73, 1205–1211.

6. Bowen, R.L.; Rodriguez, M.S. Tensile strength and modulus of elasticity of tooth structure and several restorative materials. J. Am. Dent. Assoc **1962**, 64, 378–387.

7. Peyton, F.; Mahler, D.; Hershenov, B. Elastic and mechanical properties of human dentin. J. Dent. Res **1952**, 31, 366–370.

8. Palamara, J.E.A.; Wilson, P.R.; Thomas, C.D.L.; Messer, H.H. A new imaging technique for measuring the surface strains applied to dentine. J. Dent **2000**, 28, 141–146.

9. Stanford, J.W.; Weigel, K.V.; Paffenbarger, G.C.; Sweeney, W.T. Compressive properties of hard tooth tissues and some restorative materials. J. Am. Dent. Assoc **1960**, 60, 746–756.

10. Craig, R.; Peyton, F. Elastic and mechanical properties of human dentin. J. Dent. Res **1958**, 37, 710–718.

11. Zaytsev, D.; Ivashov, A.S.; Mandra, J.V.; Panfilov, P. Deformation behavior of human enamel and dentin-enamel junction under compression. Mater. Sci. Eng. C **2014**, 41, 83–90.

12. Jantarat, J.; Palamara, J.E.; Lindner, C.; Messer, H.H. Time-dependent properties of human root dentin. Dent. Mater **2002**, 18, 486–493.

13. Rees, J.; Jacobsen, P.; Hickman, J. The elastic modulus of dentine determined by static and dynamic methods. Clin. Mater **1994**, 17, 11–15.

14. Ryou, H.; Amin, N.; Ross, A.; Eidelman, N.; Wang, D.H.; Romberg, E.; Arola, D. Young's modulus of peritubular and intertubular human dentin by nano-indentation tests. J. Mater. Sci. Mater. Med **2011**, 22, 1127–1135.

15. Han, C.F.; Wu, B.H.; Chung, C.J.; Chuang, S.F.; Li, W.L.; Lin, J.F. Stress-strain analysis for evaluating the effect of the orientation of dentin tubules on their mechanical properties and deformation behavior. J. Mech. Behav. Biomed **2012**, 12, 1–8.

16. Ziskind, D.; Fleischer, S.; Zhang, K.; Cohen, S.R.; Wagner, H.D. A novel experimental method for the local mechanical testing of human coronal dentin. Dent. Mater **2010**, 26, 179–184.

17. Ziskind, D.; Hasday, M.; Cohen, S.R.; Wagner, H.D. Young's modulus of peritubular and intertubular human dentin by nano-indentation tests. J. Struct. Biol **2011**, 174, 23–30.

18. Grimal, Q.; Raum, K.; Gerisch, A.; Laugier, P. A determination of the minimum sizes of representative volume elements for the prediction of cortical bone elastic properties. Biomech. Model. Mechanobiol **2011**, 10, 925–937.

19. Eltit, F.; Ebacher, V.; Wang, R. Inelastic deformation and microcracking process in human dentin. J. Struct. Biol **2013**, 183, 141–148.

20. Zhang, D.; Arola, D.D. Applications of digital image correlation to biological tissues. J. Biomed. Opt **2004**, 9, 691–699.

21. Hild, F.; Roux, S. CorreliQ4: A software for finite element displacement field measurements by digital image correlation. Rapp. Interne LMT Cachan **2008**, 269, 1–55.

22. Bonfield, W.; Datta, P.K. Young's modulus of compact bone. J. Biomech **1974**, 7, 147–149.

23. McElhaney, J.H. Dynamic response of bone and muscle tissue. J. Appl. Physiol **1966**, 21, 1231–1236.

24. Garberoglio, R.; Brännström, M. Scanning electron microscopic investigation of human dentinal tubules. Arch. Oral Biol **1976**, 21, 355–362.

25. Carter, D.; Hayes, W. The compressive behavior of bone as a two-phase porous structure. J. Bone Jt. Surg **1977**, 59, 954–962.

26. Kinney, J.H.; Balooch, M.; Marshall, S.J.; Marshall, G.W., Jr; Weihs, T.P. Hardness and Young's modulus of human peritubular and intertubular dentine. Arch. Oral Biol **1996**, 41, 9–13.

2

CHAPTER

FLEXURAL BEHAVIOR OF FUNCTIONALLY GRADED SANDWICH COMPOSITE

Mrityunjay R. Doddamani[1] and Satyabodh M Kulkarni[1]

[1] Mechanical Engineering, National Institute of Technology Karnataka, Surathkal, India

1. INTRODUCTION

Sandwich panels are used in a variety of engineering applications including aircraft, construction and transportation where strong, stiff and light structures are required [5]. The applicability of sandwiches could be improved if it contains a FG core which might help to distribute the stresses due to bending or in progressive absorption of energy under impact loading [6]. It is required to study the behavior of sandwich panels under these types of failures with a functionally graded material (FGM) as core to explore their new application in bullet proofing and crash worthiness.FGM's are new class of materials where property is function of geometry such as thickness, length etc [7]. These are the materials whose composition and microstructure are not uniform in space, but gradually vary following a predetermined law [8-11]. FGM's differ from composites in the sense that property is uniform in a particular direction throughout the composite. The concept of FGM's is proposed as early as 1984 by material scientists as a means of preparing thermal barrier materials [12]. Closest to FGM's is laminated composites with variation in laminate properties but they possess distinct interfaces across which properties change abruptly [13]. For example, a rocket motor casing can be made with a material system such that the inside is made of a refractory material, the outside is made of a strong metal, and the transition from the refractory material to the metal is gradual through the thickness [14]. FGM's possess a number of advantages that make them attractive in many applications, including a potential reduction of in-plane and transverse through-the-thickness stresses, an improved residual stress distribution, enhanced thermal properties, higher fracture toughness, and reduced stress intensity factors. It is worth mentioning that the distribution of the material in functionally graded structures may be designed to various spatial specifications (1). Currently, advanced processing methods to introduce compositional gradients into various material systems are being developed by materials scientists [15-17]. A typical particulate composite with prescribed variation in distribution of constituent phases could be a representative FGM.The FGM concept could be borrowed in making sandwiches with FG core which exhibit resistance (stiffness) proportional to the applied load can serve some applications better than regular sandwiches, like a spring with varying stiffness. Such a sandwich could be realized by using a particulate composite with varying volume fraction of constituents.

The flexural behavior of sandwich beams has been studied extensively by many investigators [18-23]. Studies on three point bend tests have been conducted in flexural [24-25] and short beam shear test configurations [26]. An experimental investigation of failure of piecewise FG of sandwiches subjected to three point bending is carried out by Avila[27]. In addition, fiber reinforced syntactic foams [28-30] and syntactic foam core sandwich composites have also been studied for bending properties [31]. Specific properties of sandwich with complaint FG core needs attention as it is yet to be reported.

2. OBJECTIVES AND SCOPE

From the foregoing literature survey, clear is the fact that the research reports on development of low cost materials for bullet proofing and energy absorption is hardly available. A low cost ash filled functionally graded polymer system is proposed for applications like ballistic energy absorption. The perusal of sandwich literature review prompted a thorough and systematic study on these sandwiches by performing experimental characterization of flexural properties. Therefore the work undertaken pursues the following objectives:

1. To prepare functionally graded rubber cores with varying fly ash reinforcement.

2. To plan the experiments using DOE for processing FG sandwiches with different factors (weight fraction of fly ash, core to total sandwich thickness - C/H ratio and jute skin orientation) as per L9 orthogonal array at three levels.

3. To study the effect of above parameters on mechanical properties of sandwich three point loading condition.

4. To identify the most influential factor governing the mechanical behavior of FG sandwiches.

5. To validate the gradation observed through finite element (FE) modeling using spring analogy for variations in property like uniform, linear and piecewise linear.

6. Comparison of Experimental and FE results for properties of sandwich under consideration.
7. Visual inspection of fractured FG sandwiches under different tests.

Developed FG cores are utilized in sandwiches to characterize FG sandwiches for their suitability in real world applications. Sandwiches are prepared as per design of experiments approach so that multiple factors (fly ash weight fraction, C/H ratio and jute skin orientation) at three different levels can be simultaneously analyzed. Further, these sandwiches are subjected to bending test. Another set of samples called confirmatory set is made with 25% and 35% filler by weight. Five samples are subjected to mechanical test and the response is averaged out for these five.

Furthermore, experimental values are compared with results of FE analysis. ANSYS 5.4 package is used to achieve this objective. Analysis are carried out with three gradation variations namely uniform, linear and piecewise linear. Young's modulus is computed for FG cores using FE approach and is compared with experimental result. Specific bending strength is the properties focused in simulating sandwich behavior. Finally, elaborate discussion on fractured samples is presented as the last segment of this work.

3. PROCESSING DETAILS

This section presents properties of starting material used, procedures followed for preparing FG composites and their sandwiches. Details of reagents / chemicals used at different stages like for sample curing are also described. Characteristics of the reinforcements used are also enlisted. As outlined in the objectives and scope of the work in the preceding section, the objective of the present investigation is to study the properties of functionally graded sandwiches. This section lists materials and their properties and methods adopted for processing composites with varying content of the filler.

3.1. Plan of experiment

In this work experiments are designed based on Taguchi's DOE approach for FG sandwiches [32]. Factors and levels chosen for planning the experiments for FG sandwiches are presented in Table 1. Table 2 shows orthogonal array for sandwich. Table 3 presents coding of samples bearing varying content of filler, C/H ratio and jute orientation.

Table 1. Factors and Levels selected for sandwich with FG core

Details	Wt Fraction of Fly ash % (Factor 1)	Core to thickness ratio (Factor 2)	Orientation of Jute Fabric (Factor 3)
Level 1	20	0.4	$0^0/90^0$
Level 2	30	0.6	$30^0/60^0$
Level 3	40	0.8	$45^0/45^0$

Table 2. L9 Orthogonal array for FG Sandwich

Experiment No.	Parameters		
	Weight Fraction (%)	C/H Ratio	Orientation
1	20	0.4	$0^0/90^0$
2	20	0.6	$30^0/60^0$
3	20	0.8	$45^0/45^0$
4	30	0.4	$30^0/60^0$
5	30	0.6	$45^0/45^0$
6	30	0.8	$0^0/90^0$
7	40	0.4	$45^0/45^0$
8	40	0.6	$0^0/90^0$
9	40	0.8	$30^0/60^0$

Table 3. Description of sample codes used for sandwiches

Sample code	Description
$W_aR_bO_c$	Sandwich specification
W	Indicates factor 1 (Wt. fraction of fly ash)
a	Levels of factor 1 in % (20, 30, 40)
R	Indicates factor 2 (C/H ratio)
b	Levels of factor 2 (0.4, 0.6, 0.8)
O	Indicates factor 3 (Fiber Orientation in skin)
c	Levels of factor 3 (0°/90°, 30°/60°, 45°/45°)

Experimentation is done with due considerations to all the above parameters with both configurations of gradation namely rubber up and ash up. In each trial minimum of five replicates are tested. Average of the measured parameters for each set of replicates is subjected to statistical ANOVA to find the most influential factor governing the behavior using Minitab release 14 statistical analysis tool.

3.2. Materials

Details of materials used for main constituents of sandwiches (core and skin) are presented hereafter.

3.2.1. Core for FG sandwich

From the standpoint of cost, availability,and the scarce literature prompted for going in for an elastomeric material which is naturally occurringand known by the name 'natural rubber' as the matrix material. Further it is reinforced with fly ash and is used as core in sandwich.

As many of the polymeric systems for developing FGM's are generally with the tag of expensiveness associated, it is decided to examine the gradation in composition and its subsequent mechanical behavior when an abundantly available lower density possessing fly ashes are the filler materials for the core. Fly ashes are fine particulate waste products derived during generation of power in a thermal power plant. These have aspect ratios closer to unity and hence are expected to display near isotropic characteristics. These inexpensive and possessing good mechanical properties, when used with well established matrix systems help to reduce the cost of the system and at the same time either retain or improve specific and desirable mechanical properties. Fly ash has attracted interest [33-34] lately, because of the abundance in terms of the volume of the material generated and the environmental-linked problems in the subsequent disposal. Fly ash mainly consists of alumina and silica, which are expected to improve the composite properties. Fly ash also consists to some extent hollow spherical particles termed as cenosphere which aid in maintenance of lower density values for the composite, a feature of considerable significance in weight-specific applications [35-36]. Again, as the fillers do not come under irregular shape, the resin spread, is better and as the ashes are essentially a mixture of solid, hollow and composite particles displaying near isotropic properties, developing newer and utilitarian systems using them should be an interesting and challenging task [37]. Compositional details of a fly ash particle are tabulated in Table 4.

Table 4. Compositional details of fly ash particle

Constituent	Wt. %
SiO_2	63
Al_2O_3	26.55
CaO	0.42
Fe_2O_3	6.7
TiO_2	2.47

3.2.2. Skin used in sandwich

Further on, in this effort, for the skins too, it is decided to employ instead of the well explored man-made fibers like glass, carbon or aramid a fairly strong but naturally occurring one going by the name 'jute fiber' andknown for its inexpensiveness. Jute is an attractive natural fiber for use as reinforcement in composite because of its low cost, renewable nature and much lower energy requirement for processing. In comparison to glass fibers jute has higher specific modulus and lower specific gravity as against that of glass fiber. Jute reinforced plastics offer attractive propositions for cost-effective applications [38]. These in the form of laminates have much better properties than their neat resin counterparts [39]. Better properties of woven jute fabric reinforced composites demonstrated their potential for use in a number of consumable goods in an earlier literature [40]. Substantial increases in flexural modulus and strength with small amounts of reinforcement of unidirectional jute have also been reported [41]. Keeping these things in mind a bi-directionally woven jute fabric is used in different orientations. Table 5 gives the brief overview of comparison between glass fibers and jute fibers.

Table 5. Mechanical Properties of Glass and Jute Fibers

Property	E-glass	Jute
Specific Gravity	2.5	1.3
Tensile Strength (MN/m^2)	3400	442
Young's Modulus (MN/m^2)	72	55.5
Specific Strength (MN/m^2)	1360	340
Specific Modulus (GN/m^2)	28.8	42.7

The major drawback of natural fiber reinforced composites is due to its affinity towards moisture. Many experimental studies have shown that compatible coupling agents are capable of either slowing down or preventing the de-bonding process and hence moisture absorption even under severe environmental conditions such as exposure to boiling water. Jute fibers/fabrics can be modified chemically through graft co-polymerization and through incorporation of different resin systems by different approaches.

3.2.3. Matrix for skin

For fabricating both the skins and core a matrix system is required. A thermosetting epoxy is chosen for this purpose as far as the skins are concerned. The adhesive used in present work consists of a medium viscosity epoxy resin (LAPOX L-12) and a room temperature curing polyamine hardener (K-6) supplied by ATUL India Ltd. Epoxy resin is selected as the material for the matrix system because of its wide application, good mechanical properties, excellent corrosion resistance and ease of processing. Some details including density of the constituents of the matrix system chosen are listed in Table 6.

Table 6. Details of the constituents of matrix used for skin in sandwich

Constituent	Trade name	Chemical name	Epoxide equivalent	Density (kg/m^3)	Supplier	Parts by weight
Resin	LAPOX L-12	Diglycidyl Ether of bisphenol A (DGEBA)	182 - 192	1162	ATUL India Ltd.	100
Hardener	K-6	Tri ethylene Tetra amine (TETA)	----	954	- do -	10-12*

With these materials in hand, FG sandwiches are prepared for mechanical testing.

4. PROCESSING OF FG SANDWICH

FG cores used in the present work are produced using the following procedure. The gradation in the core is expected due to differential settling of the particles with different densities at different depths in the rubber matrix. A measured quantity of natural latex is mixed with pre-weighed amounts of fly ash, sulphur (vulcanizer) and zinc oxide (catalyst) [42] by adopting gentle stirring for about 1 hour. The mold employed for preparation of core specimen is completely covered on all sides with teflon sheet. Subsequently, silicone releasing agent is applied to facilitate ease of removal of the cast sample at a later stage. The mixture is then slowly decanted into the mold cavity followed by curing at 90°C in an oven for about 5-6 hours. The cured rigid plate sample is withdrawn from the mold and the edges trimmed. Figure 1 presents one such FG sample which in turn will be used as core in sandwiches.

As regards the sandwich skins, a bi-directional woven jute fabric procured from M/S Barde Agencies, Belgaum, Karnataka, India is used. This fabric is cut into layers of dimensions depending on the sandwich sample size in required orientation. Thickness of each fabric piece is 0.5 mm. All the layers of jute fabric are heated in an oven at 70^{0}C for 5-10 minutes to remove moisture present. The jute stack thickness to form the thin skin, on either side of FG core, is computed. This enables one to arrive at the required number of fabric layers to be used, as thickness of each layer is known. Based on required C/H ratio number of fabric layers to be used are determined (Table 7).

Figure 1. Functionally graded core sample drawn out of mold

Table 7. Jute layer arrangement for achieving C/H ratios in sandwich

C/H Ratio	Core thickness - C (mm)	Number of jute layers below core	Number of jute layers above core	Sandwich thickness - H (mm)
0.4	4	6	6	10
0.6	6	4	4	10
0.8	8	2	2	10

With this background data on hand to begin with, the required fabric pieces are dipped in mixture of epoxy and K-6 hardener and placed on base plate forming the bottom stack of the sandwich.Now, the earlier mentioned procedure-wise made FG core dipped in resin mixture is placed on the bottom stack of skins. Finally, over such an arrangement, the remaining layers of jute fabrics having undergone the same procedure for fabrication are stacked to constitute the top skin. A procedure of this nature should help in ensuring a greater degree of spread of the resin on the fibrillar jute.Following this, the excess resin is made to come out by a squeezing operation that is aided by tightening of the mold top plate. The mold assembly is then cured at room temperature for about 24-26 hours. The sandwich sample is withdrawn from the mold and trimmed to the required size. Similarly numbers of samples are made with various core thickness and orientation in skin as schematically illustrated in Figure 2. Figure 2 (a) shows top view with different orientations and while the front view with varying core thickness to total sandwich thickness (C/H ratio) is presented in Figure 2 (b).

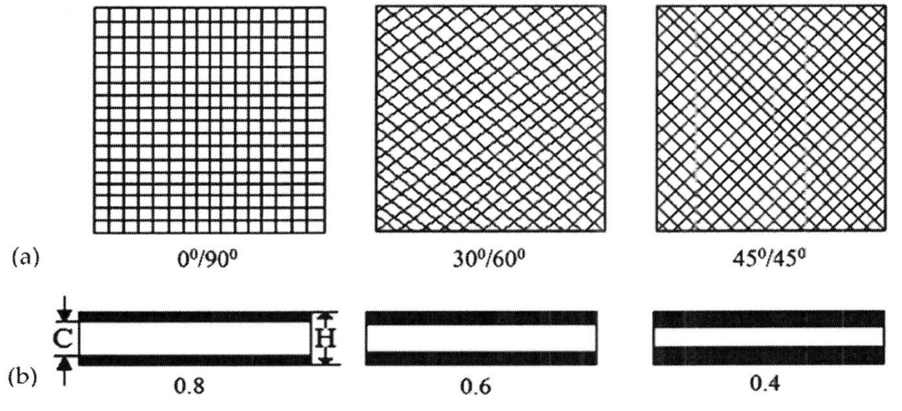

Figure 2. a). Orientation of jute strands in the sandwich skins, (b). Variation of C/H ratio considered for analysis

5. EXPERIMENTAL DETAILS

The mechanical testing of sandwich composites to obtain parameters such as strength, stiffness etc. is a time consuming and often difficult process. It is, however, an essential process, and can be somewhat simplified by the testing of simple structures such as flat coupons. The data obtained from these tests can then be directly related with varying degrees of simplicity and accuracy to any structural shape. The test methods outlined in this section merely represent a small selection available to the composites scientist. Various FG sandwiches fabricated are characterized for three point bending condition. Influence of rubber up (rubber rich region towards the top) and ash up (ash rich region below the loading point) configurations are critically analyzed. Expected gradation in FG cores is presented in Figure 3 (rubber up and ash up).

The three point bending test is carried out in accordance with ASTM C 393 [42] using Instron universal testing machine of model 4206 with loading capacity ranging from 0.1 N to 150 kN. Figure 4 shows the sandwich sample mounted on flexural test set-up. The thickness to span ratio of the tested sandwich samples is 1:16. The crosshead displacement rate is maintained at 2 mm/min. The load deflection data is recorded at equal intervals up to a point at which the specimen shows the first sign of failure.

From load deflection data, bending modulus and strength are estimated using relations 1 and 2 respectively and the mean of five samples in each sandwich configuration is used for inference.

$$\text{Specific bendingmodulus} = \frac{\text{Flexural modulus}}{\text{Weight density}} = \frac{E_{bending}}{\rho \text{x}\ g} \quad (1)$$

$$\text{Specific bending strength} = \frac{\text{Ultimate strength}}{\text{Weight density}} = \frac{\sigma_{u_{bending}}}{(\rho x g)} \quad (2)$$

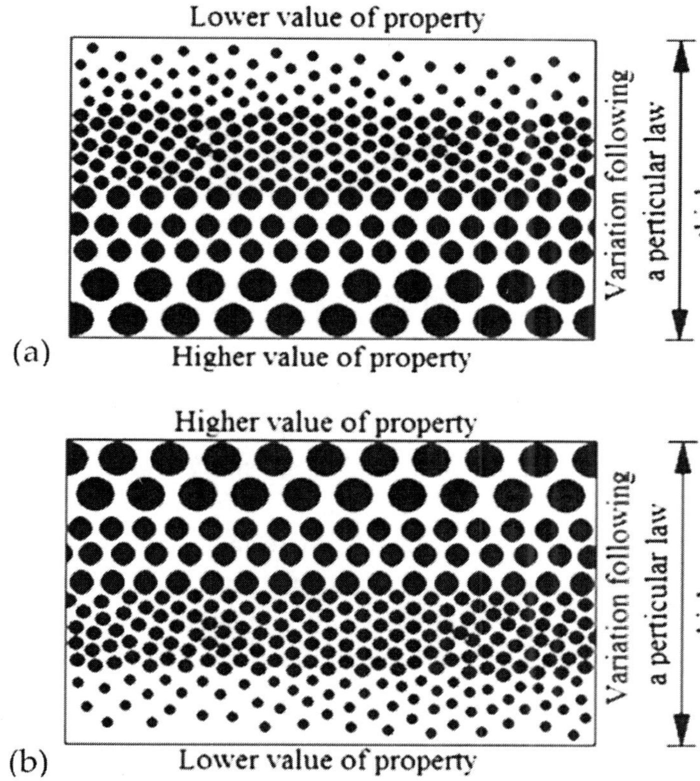

Figure 3. a). Rubber Up condition in FG core, (b). Ash Up condition in FG core

Figure 4. Sandwich sample mounted on flexural test set-up

where

$$\sigma_{u_{bending}} = \frac{6M}{BH^3} \text{ and}$$

and

$$M = \frac{FL}{4} \tag{3}$$

5.1. Details of finite element modelling

As outlined earlier, FE model helps to model the constituents of the FG composites and their sandwiches to study the interactions of these in load transfer and mechanisms influencing their failure. To understand and predict the effect of material as well as geometrical parameters on the mechanical behavior of FG fly ash filled rubber composites and their sandwiches finite element analysis can be a very effective technique. Towards this, a simple disctretized model is built in the software ANSYS® representing FG composites with properties varying from top layer to bottom representing gradation.

Static analysis is performed using FEM software ANSYS 5.4. In this analysis a two dimensional model of a FG system is constructed and meshed with 4-node PLANE42 element. Three different mesh sizes are tested with 4-node elements to check the convergence of the model, based on which medium mesh size (element edge length is taken as 0.5) is selected. Number of nodes and elements used in the analysis are 800 and 5000 respectively.

Finite element values are compared with experimental ones for bending behavior of FG sandwich. At the contact surfaces of the layers and between layers and faces of sandwich glue conditions are applied to eliminate relative movement of layers with respect of each other. Furthermore, nodes are merged at the interface allowing proper coupling between layers and interfaces. Figure 5 shows finite element mesh with boundary conditions as a typical case considered for three point bending analysis. Skins are being represented by top and bottom portions of the structure whereas in between are the four layers having graded properties.

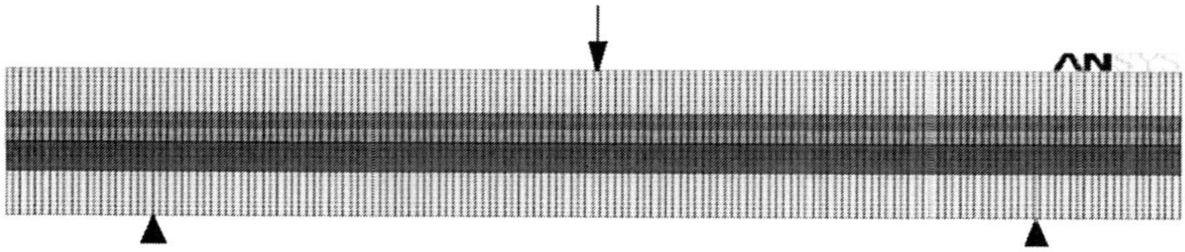

Figure 5. Finite element mesh with boundary condition for FG sandwich

While modeling gradation in ANSYS 5.4, the analogy of springs is used having differing stiffness ($K_1 < K_2 < K_3 < K_4$) from the top layer to bottom (Figure 6).

Sandwiches with FG core are modeled in FEA package ANSYS 5.4 [43] as emphasized before. Three different gradations of filler U (uniform), L (linear) and PL (piecewise linear) are considered during modeling of FG cores (Figure 7). Young's modulus and density of FG cores are determined for different weight fractions of fly ash from constituent properties are provided as input to FEA (Table 8).

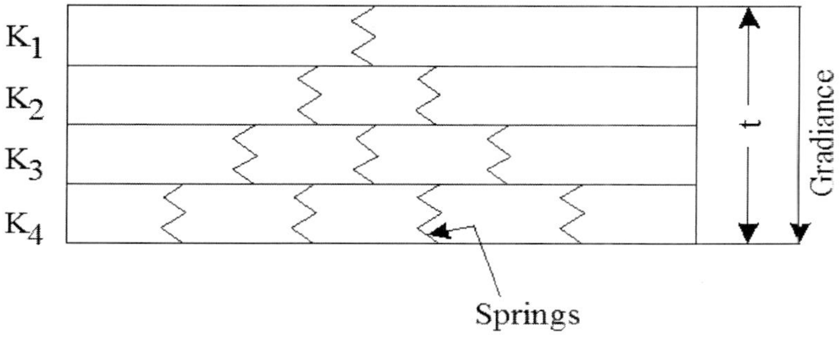

Figure 6. Spring analogy for gradation in modulus of core material

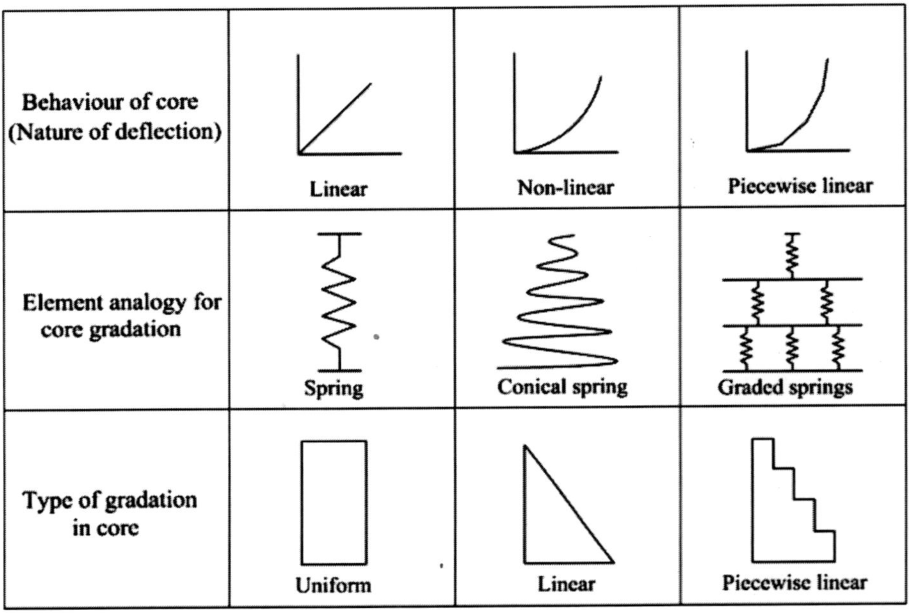

Figure 7. FG rubber core configurations used in FEA

Fly ash distributions taken into account for uniform configuration are 20%, 30% and 40% through the thickness. For these weight fractions Young's modulus is estimated using inverse rule of mixtures For skins, young's modulus is estimated by preparing five tensile samples of jute/epoxy with orientations of $0^0/90^0$, $30^0/60^0$ and $45^0/45^0$ which are subsequently tested as per ASTM D3039 [44] guidelines. Density of skins is determined experimentally using procedure outlined in ASTM D792 [45]. Table 8 presents properties of core and skin used in the FE analysis. Results of FE analysis are compared with experimental values.

Bending tested samples are subjected to visual observation using regular photography technique for FG sandwich. These methods came in handy during the characterization of failures especially in impact failed samples.

Table 8. Core and skin properties used in FEA

FG Core							
Wt. % of fly ash	Young's modulus (GPa)			Density (Kg/m³)			Element
	U	L	PL*	U	L	PL*	
20%	0.7575	0.65 (upper)	0.65 (L1)	1168.4	1163.9 (L1)	1162.8 (L1)	2D Plane 42
		0.75 (middle)	0.71 (L2)		1167.5 (L2)	1165.2 (L2)	
		0.88 (bottom)	0.79 (L3)		1172.5 (L3)	1168.2 (L3)	
			0.88 (L4)			1173.5 (L4)	
30%	0.89	0.68 (upper)	0.68 (L1)	1330.2	1324.5 (L1)	1323.9 (L1)	
		0.865 (middle)	0.79 (L2)		1331.1 (L2)	1328.4 (L2)	
		1.15 (bottom)	0.94 (L3)		1336.9 (L3)	1334.6 (L3)	
			1.15 (L4)			1337.2 (L4)	
40%	1.1	0.71 (upper)	0.71 (L1)	1444.7	1435.2 (L1)	1434.9 (L1)	
		1.015 (middle)	0.88 (L2)		1445.7 (L2)	1442.8 (L2)	
		1.66 (bottom)	1.15 (L3)		1452.6 (L3)	1450.6 (L3)	
			1.66 (L4)			1455.2 (L4)	

Jute / Epoxy skin			
Orientation	Ex (GPa)	Ey (GPa)	Density (Kg/m³)
0⁰/90⁰	3.25	2.5	1468
30⁰/60⁰	1.63	1.25	1451.2
45⁰/45⁰	2.29	1.77	1444.3

6. RESULTS AND DISCUSSION

FG sandwiches are tested for Density, the results of which are presented in Table 9.

Experimental density values are subjected to statistical analysis (MINITAB 14) to propose regression equation which is presented in equation 3.

$$\text{Density } (\text{Kg/m}^3) = 1099 + [11.6 \times \text{Fly ash weight }\%] + [29.7 \times \text{C/H Ratio}] - [0.459 \times \text{Jute Orientation}] \quad (4)$$

Table 9. Density results of FG sandwiches

Sandwich code	Trial-1	Trial-2	Trial-3	Trial-4	Trial-5	Density (Kg/m^3)
$W_{20}R_{0.4}O_0$	1325.6	1328.9	1329.4	1332.8	1330.8	1329.5
$W_{20}R_{0.6}O_{30}$	1333.5	1334.8	1336.2	1336.4	1331.6	1334.5
$W_{20}R_{0.8}O_{45}$	1342.8	1350.7	1348.6	1345.5	1348.9	1347.3
$W_{30}R_{0.4}O_{30}$	1465.8	1464.6	1460.3	1462.1	1463.2	1463.2
$W_{30}R_{0.6}O_{45}$	1435.2	1435.9	1431.9	1432.8	1433.7	1433.9
$W_{30}R_{0.8}O_0$	1467.1	1466.9	1469.3	1470.5	1467.2	1468.2
$W_{40}R_{0.4}O_{45}$	1547.6	1549.8	1551.7	1550.6	1548.8	1549.7
$W_{40}R_{0.6}O_0$	1599.5	1598.8	1595.6	1594.4	1596.2	1596.9
$W_{40}R_{0.8}O_{30}$	1564.1	1561.8	1562.4	1560.9	1563.8	1562.6

Equation 3 comes handy, which predicts density for large number of samples with varying combination of factors within the range of chosen levels without experimentation. Density increases with filler content as well as with C/H ratio (core to thickness ratio) being positive coefficients while shows a decreasing trend with increase in jute orientation. Obvious reason for this might be lower specific weight with increasing skin orientation.

Three point bending behavior of a FG sandwich composite is investigated under flexural loading condition. Results are analyzed for specific modulus and specific bending strength. Load deflection data is traced all along the path. The load and corresponding deflection data is noted at equal intervals up to a maximum load at which the specimen shows the first sign of failure (point 'A'). The load and deflections obtained during testing are plotted. A typical load deflection curve is shown in Figure 8.

Load-displacement consists of an initial linear part followed by a nonlinear portion (Figure 8). A nonlinear mechanics of materials analysis that accounts for the combined effect of the nonlinear behavior of the facings and core materials (material nonlinearity) and the large deflections of the beam (geometric nonlinearity) are observed. The nonlinear load-deflection behavior of the beams is attributed to the combined effect of material and geometric nonlinearity. The material nonlinearity of the sandwich beam is due to the nonlinear normal stress-strain behavior of the facing material and the FG core. For long beam spans, even though there is a geometric nonlinearity effect, the overall load-deflection curve of the beam does not deviate much from linearity.

For long beam spans the nonlinearity of the load-deflection curve is mainly due to the combined effect of the facings nonlinearity and the large deflections of the beam. Both effects, however, have a small contribution to the load-deflection behavior, which shows a small deviation from linearity. Some of the general observations made are listed below.

1. The load decreases sharply after the end of the elastic region due to failure initiation in sandwich composites (A to B).
2. All samples have shown small linear region (B to C) before skin failure in compressive side.
3. Variation in displacement value at which peak load is observed for various types of FG sandwiches is considerable.
4. The failure originates on the tensile side.

6.1. Specific bending modulus

From load deflection data the average specific modulus and strength for five samples (Table 10) are estimated using equations 1 and 2.

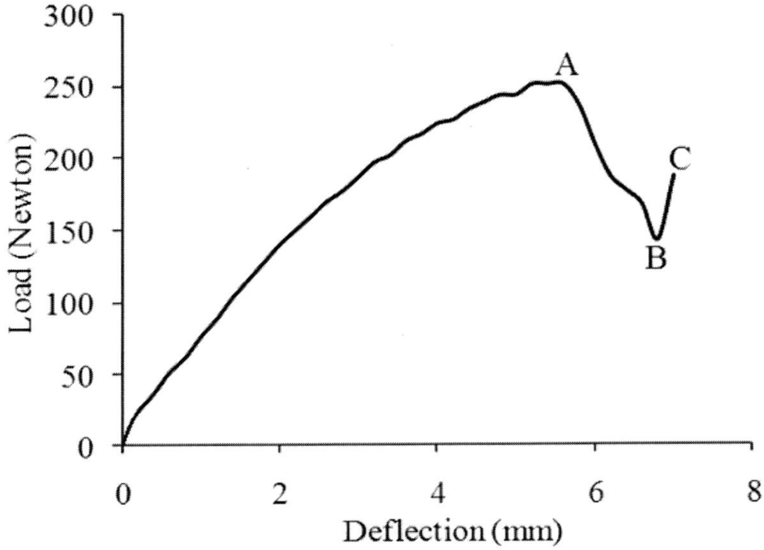

Figure 8. Load-deflection behavior under three point bend test for sandwich

It can be clearly seen from the table that, rubber up configuration registered higher results compared to ash up condition for both the properties in the range of 7 to 30%. Constrained straining and resisting forces set up in the FG core might be the reasons for such an observation in bending test as depicted in Figure 9.

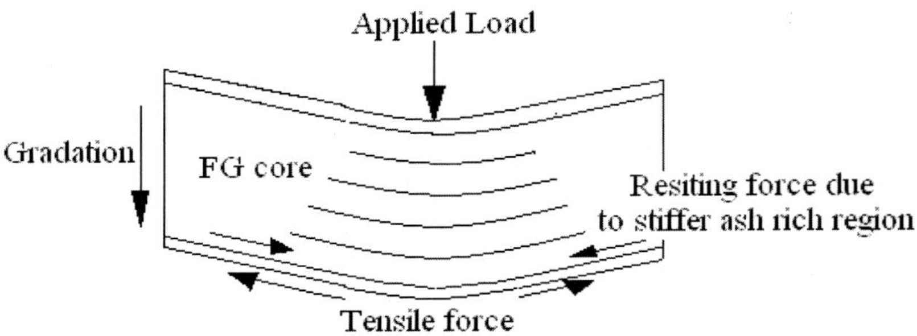

Figure 9. Loads acting on FG sandwich in bending test

Table 10. Specific bending modulus and strength for FG sandwich

Sample coding	Sp. flexural modulus (MPa/Nm^{-1})				Sp. flexural strength (MPa/Nm^{-1})			
	Rubber Up	Avg.	Ash Up	Avg.	Rubber Up	Avg.	Ash Up	Avg.
$W_{20}R_{0.4}O_0$	3945.23	3953.07	3410.9	3410.29	132.7	128.1	103.55	98.81
	3933.7		3404.2		128.1		105.32	
	3961.5		3419.16		127.7		95.47	
	3963.7		3416.4		128.1		99.1	
	3961.2		3400.8		123.9		90.61	
$W_{20}R_{0.6}O_{30}$	5322.6	5319.4	4540.15	4545.36	85.3	88.1	72.3	70.7
	5306.8		4545.39		88.1		70.7	
	5321.3		4539.4		83.5		71.9	
	5322.9		4544.95		92.4		70.7	
	5323.4		4556.9		91.2		67.9	
$W_{20}R_{0.8}O_{45}$	7391.4	7387.91	6150.4	6155.14	54.59	54.59	45.23	48.75
	7387.91		6160.73		54.5		50.4	
	7377.4		6155.14		57.7		47.9	
	7393.4		6155.14		58.1		46.8	
	7389.42		6154.28		48.06		53.42	
$W_{30}R_{0.4}O_{30}$	2996.2	3001.3	2390.31	2398.92	141.4	141.4	115.43	113.26
	3003.1		2398.92		141.4		116.23	
	3004.5		2398.92		145.5		113.26	
	3001.3		2398.92		149.1		111.59	
	3001.4		2407.53		129.6		109.79	
$W_{30}R_{0.6}O_{45}$	4043.3	4045.36	3533.59	3533.57	94.76	101.23	80.17	78.75
	4047.6		3528.61		99.14		81.34	
	4042.4		3531.75		95.35		75.46	
	4045.36		3523.16		106.4		79.1	
	4048.12		3550.73		110.5		77.68	
$W_{30}R_{0.8}O_0$	6559.3	6562.65	6018.2	6018.2	148.7	153.1	120.1	119.3
	6562.8		6018.2		149.2		121.3	
	6570.4		6018.2		157.3		118.3	
	6560.55		6020.9		151.2		121.54	
	6560.21		6015.5		159.1		115.26	
$W_{40}R_{0.4}O_{45}$	2134.3	2138.92	1702	1692.71	149.3	151.4	110.34	117.17
	2138.69		1692.67		148.7		121.56	
	2141.92		1688.2		152.4		117.17	
	2139.26		1690.4		159.3		120.23	
	2140.42		1690.3		147.3		116.55	
$W_{40}R_{0.6}O_0$	4372.5	4365.98	4060.12	4065.98	188.98	192.21	159.21	154.45
	4370.28		4068.63		193.5		155.29	
	4365.39		4065.98		199.7		152.8	
	4360.87		4059.3		191.49		155.7	
	4360.86		4075.86		187.38		149.25	
$W_{40}R_{0.8}O_{30}$	6515.5	6518.2	6050.3	6062.65	155.23	159.53	121.44	125.45
	6520.7		6070.4		151.8		123.3	
	6521.4		6060.9		161.32		127.56	
	6518.2		6058.5		164.2		128.9	
	6515.2		6073.15		165.1		126.05	

Rubber up condition of FG core in sandwich represents ash rich region on tensile side. Crack initiation is observed to be from tensile region to compressive region in pre sent loading case. In rubber up condition, as stiffer zone is near tensile region, sandwich can take up higher loads resulting in better performance compared to homogenous cores and ash up condition in FG core. Thereby, such sandwiches are excellent examples of optimized designs.

Developed FG sandwiches can be used in practical cases wherein structures are continuously subjected to bending loads. Depending upon whether load is acting downwards or upwards sandwiches can be suitable placed with either rubber up or ash up configuration as regards to FG cores.

Figure 10 shows the signal to noise (SN) response plot for specific bending modulus with respect to the parameters under study. Response of SN ratio in Specific bending modulus for Rubber Up condition is presented in Table 11.

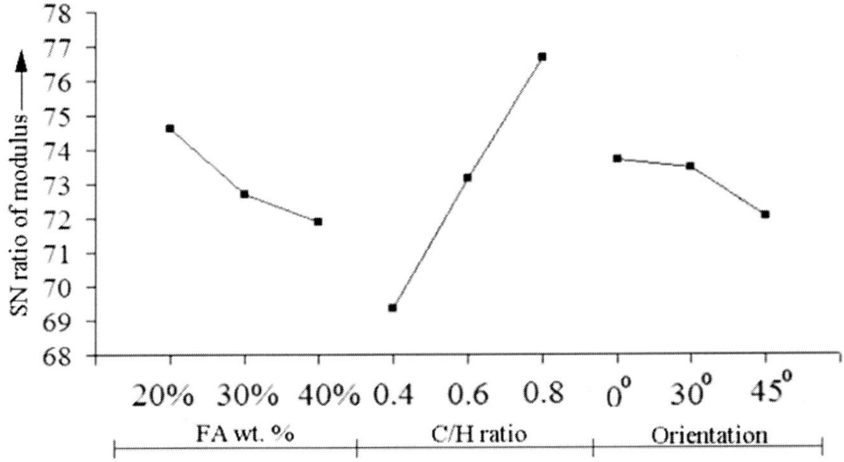

Figure 10. Variation of SN ratio in specific bending modulus (Rubber Up)

Table 11. SN ratio table for specific bending modulus (Rubber Up)

	Fly ash weight %	C/H ratio	Orientation
Level 1	74.61	69.36	73.69
Level 2	72.68	73.15	73.45
Level 3	71.90	76.66	72.04
Effect	2.71	7.30	1.66
Rank	2	1	3

From the data analysis, vide response Table 11, it is seen that C/H ratio and fly ash % exhibit greater influence compared to the orientation. It is further observed from the Table and Figure 10 that samples with fly ash content of 20%, C/H of 0.8 and an orientation of $0^0/90^0$ possess highest specific bending modulus. This could be due to higher C/H ratio implying larger rubber rich region imparting higher modulus to sandwich system.

6.2. Specific bending strength

Results of specific bending strength from Table 10 are statistically analyzed and are used to rank the variables as presented in Table 12.

Table 12. SN ratio table for specific bending strength (Rubber Up)

	Fly ash weight %	C/H ratio	Orientation
Level 1	38.6	42.92	43.84
Level 2	42.27	41.56	41.99
Level 3	44.44	40.83	39.48
Effect	5.85	2.09	4.36
Rank	1	3	2

From SN response Table, it can be seen that specific bending strength behavior is prominently governed by fly ash weight % followed by orientation and C/H ratio. Figure 11 presents SN plot for specific bending strength incase of rubber up condition.

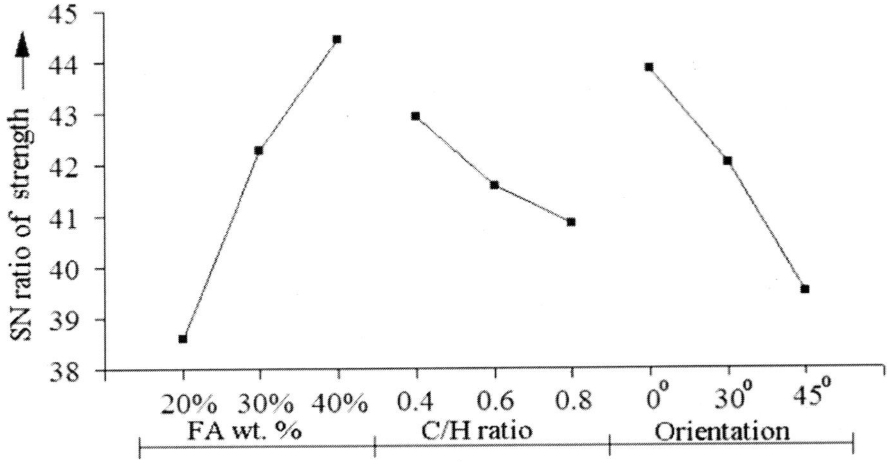

Figure 11. Plot of SN ratio in specific bending strength (Rubber Up)

From SN response plot shown in Figure 11, the best combination for specific strength is a sample with fly ash content of 40%, C/H of 0.4 and orientation of $0^0/90^0$. Reasons for this could be stiffening effect due to high modulus filler and larger skin-epoxy component for lower C/H ratios. Similar results are observed for ash up configuration. Even though $W_{20}R_{0.8}O_{45}$ and $W_{40}R_{0.6}O_0$ are showing higher values (Table 10) for modulus and strength respectively, inference on basis of these will not lead to an appropriate conclusion. The reason being these values are merely based on average of means. Inference on the grounds of SN analysis leads to a meaningful conclusion as it takes means and data spread into account. By the SN ratio analysis the best sandwich configurations are $W_{20}R_{0.8}O_0$ and $W_{40}R_{0.4}O_0$ for specific modulus and strength respectively. Similar observation is noted for ash up configuration. Regression equation is proposed based on the experimental data for specific bending properties are presented in equations 4-7.

$$\text{Specific Bending Modulus (Rubber Up)} = 1151 - [60.6 \times \text{Weight \% of Fly ash}] + [480 \times \text{C/H Ratio}] - [8.38 \times \text{Jute Orientation}] \quad (5)$$

$$\text{Specific Bending Strength (Rubber Up)} = 70.4 + [3.87 \times \text{Weight \% of Fly ash}] - [44.7 \times \text{C/H Ratio}] - [1.19 \times \text{Jute Orientation}] \quad (6)$$

$$\text{Specific Bending Modulus (Ash Up)} = 342 - [38.2 \times \text{Weight \% of Fly ash}] + [8945 \times \text{C/H Ratio}] - [14.2 \times \text{Jute Orientation}] \quad (7)$$

$$\text{Specific Bending Strength (Ash Up)} = 54.2 + [2.98 \times \text{Weight \% of Fly ash}] - [29.8 \times \text{C/H Ratio}] - [0.912 \times \text{Jute Orientation}] \quad (8)$$

6.3. Finite element analysis

Specific bending strength is estimated by simulating the sample and loading (Gupta et al. 2008) in FEA. Figure 12 represents the plot for bending stress in the sample for one typical loading case.

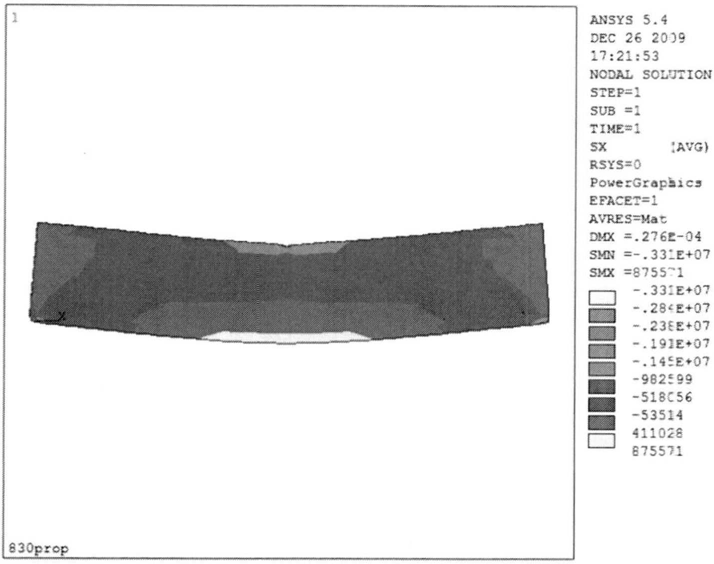

Figure 12. Bending stress in x-direction for typical case in FG sandwich

The breaking load taken from experiment is applied on FE model. For this applied load, maximum stress (von misses criteria) is recorded and finally specific strength is determined by taking the ratio of maximum stress to the weight of sample. The specific strength values obtained from FEA for three variations in gradation (Uniform-U, Linear-L and Piecewise linear-PL) and with experimental approach is presented in Table 13.

Table 13. Specific bending strength (MPa/Nm^{-1}) results for sandwich

Sandwich configuration	FEA			Experimental	% Error with PL
	U	L	PL		
$W_{20}R_{0.4}O_0$	115.4	119.6	132.75	128.1	3.50
$W_{20}R_{0.6}O_{30}$	78.2	81.5	92.58	88.1	4.84
$W_{20}R_{0.8}O_{45}$	46.9	48.9	58.58	54.59	6.81
$W_{30}R_{0.4}O_{30}$	125.1	130.2	147.34	141.4	4.03
$W_{30}R_{0.6}O_{45}$	84.4	88.8	110.38	101.23	8.29
$W_{30}R_{0.8}O_0$	129.7	137.6	160.88	153.1	4.84
$W_{40}R_{0.4}O_{45}$	126.6	131.3	169.11	151.4	10.47
$W_{40}R_{0.6}O_0$	175.2	179.5	201.42	192.21	4.57
$W_{40}R_{0.8}O_{30}$	140.2	145.6	165.7	159.53	3.72

It is significant to note that the experimental results for specific bending strength match well with FEA values especially for the ones with PL gradation. It is observed that bending strength obtained from FEA is slightly higher than experimental values. This could be due to inability of modeling inhomogenities creeping in during the processing of samples which may result in lowering specific strength.

6.4. Discussion on fractured samples

Within the elastic region of the load-displacement curve (Figure 8), where no damage is induced, the responses of all specimens to the applied loads are quite similar. This is visible in the form of nearly constant slope in the elastic region of the load-displacement curves.

It is observed that the failure starts in the form of crack origination on the tensile side of the specimen as displacement increases. On further loading, the skin of the sandwich composite that is on the tensile side tends to fracture, causing the final failure of the specimen. However, it is not significant enough to lead to the final failure of the specimen. It is observed that the entire specimen fractures at a much later instant of skin fracture. Appearance of small linear region (B to C in Figure 8) at the end in the load-displacement curves is due to stiffening of FG core before final failure. During the loading process, deformation also takes place in the compression side of the specimen. Cracks initiate from the tensile side and propagate to the compressive side within the core in all sandwiches.

It is worth discussing the mode of failure. Sandwich samples tested under bending did not display the distinct separation into pieces at failure. The FG core being compliant is observed to be successfully absorbing media.Basically two types of failure mechanisms observed are skin cracking and delamination between skins and core. Figure 13 shows the failed sandwich specimens with their failure modes.

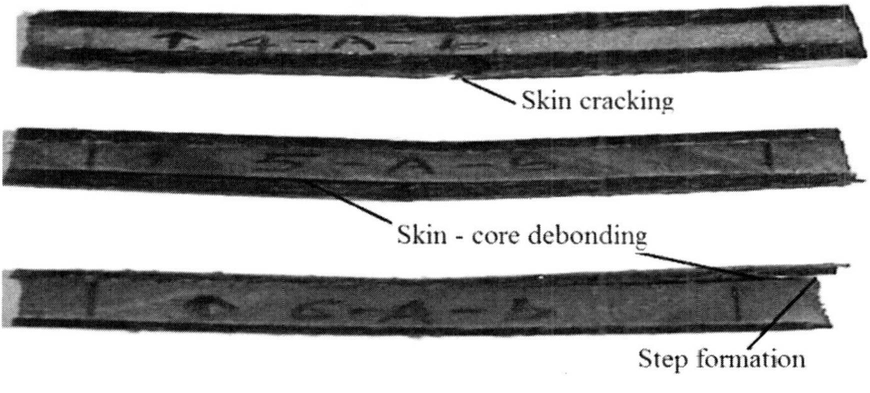

Figure 13. Sandwich failure modes under three point bending loads

The sandwich beams failed at the center of the two supporting rollers. In this portion of the beam, the shear force is zero and only the pure bending exists. Thus, the sandwich samples are capable of resisting higher bending moment. As the load on the specimen is increased, failures first start under the loads in tensile region and then they propagate towards the compressive zone through compliant FG core. All the samples failed under skin tension or compression and skin - core debonding. The sandwiches with higher C/H ratio have shown skin - core debonding. FG core takes up most of the load applied for higher C/H ratios (lesser skin thickness). Since core is made up of rubber composite being compliant in nature, relative movements are set up with respect to skin resulting in inter laminar shear stresses. As magnitude of these stresses crosses the adhesive strength delamination creeps in.Some sandwich samples are seen to be intact even after the first sign of failure. These samples exhibited a spring back effect. Samples bearing lower C/H ratio have failed mainly because of skin cracking along the jute orientation. Few samples failed due to shearing at skin-core interface displayed step formation.

7. CONCLUSIONS

This section highlights the significant conclusions drawn from the results presented earlier. Major inferences from both experimental and finite element investigations are discussed below.

Density of FG sandwiches increases with filler content and C/H ratio while decreases with jute orientation. An experimental investigation of sandwiches under bending loads for specific modulus and specific strength shows that C/H ratio and fly ash weight fraction are the influential factors respectively. Specific bending modulus in both cases (i.e. rubber up and ash up) the sample $W_{20}R_{0.8}O_0$ registered the higher value while $W_{40}R_{0.4}O_0$ shows higher value of specific strength.

Rubber up configuration registered higher results compared to ash up condition for modulus and strength. The ash up condition recorded about 30% increase in strength. Increasing fly ash weight fraction rendered an increase in bending strength of about 29% for rubber up condition. Specific strength values estimated from FEA for bending loads match well with experimental results especially for piecewise gradation.

REFERENCES

1. Nicoleta Alina Apetre.Sandwich Panels with Functionally Graded CorePhD thesis.Graduate school of the university of Florida; 2005

2. J. Vinson, The Behavior of Sandwich Structures of Isotropic and Composite MaterialsPennsylvania: Technomic; 1999

3. Gdoutos EE, Daniel IM.Failure modes of composite sandwich BeamsJournal of Theoretical and Applied Mechanics 2008

4. Venkata Dinesh Muthyala.Composite sandwich structure with grid stiffened coreMaster thesis. Graduate Faculty of the State University and Agricultural and Mechanical College, Louisiana, 2007

5. D. Zenkert, Handbook of Sandwich Construction. London:EMAS, Chameleon Press; 1997

6. S. Abrate, Impact on Composite Structures Cambridge: Cambridge University Press; 1998

7. Paulino GH, Jin ZH.Correspondence Principle in Viscoelastic Functionally Graded MaterialsJournal of Applied Mechanics 2001129 EOF131 EOF

8. M. Koizumi, of. F. G. M. Concept, Ceramic, Transactions 199334310

9. S. Suresh, A. Mortensen, Fundamentals of Functionally Graded Materials. London: Institute of Materials; 1998.

10. Miyamoto Y, Kaysser WA, Rabin BH, Kawasaki A, Ford RG. Functionally Graded Materials: Design, Processing and Applications.Dordrecht:Kluwer Academic; 1999.

11. V. Cannillo, T. Manfredini, C. Siligardi, A. Sola, Preparation and experimental characterization of glass-alumina functionally graded materials.Journal of the European ceramic society 2006

12. Al-Ajmi MA, Alhazza KA.th International Conference on Composite Science and Technology, ICCST/5 2005conference proceeding. Vibration and damping analysis of sandwich beams with functionally graded viscoelastic core. Sharjah, United Arab Emirates, 125128

13. V. Parameswaran, A. Shukla, Processing and Characterization of a Model Functionally Gradient Material.Journal of Materials Science 2000

14. S. Venkataraman, B. V. Sankar, A. I. A. A. A. S. M. E. A. S. C. E. A. H. S. A. S. C. 42nd, Structural. Structures, Dynamics, Conference. Materials, 2001conference proceeding.Analysis of sandwich beams with functionally graded core.AIAA-20011281Seattle, Washington.

15. Kitey. R. Kirigulige, H. V. Tippur, Dynamic fracture behaviour of model sandwich structures with functionally graded core: a feasibility study. Composites Science and Technology 2005

16. A. Pollien, Y. Conde, L. Pambaguian, A. Mortensen, Graded open-cell aluminium foam core sandwich beamsMaterials Science and Engineering: A 20059 EOF

17. N. A. Gupta, graded. functionally, foam. syntactic, for. material, energy. high, under. absorption, compression, Materials Letters2007979 EOF

18. D. Krajcinovic, Beam. Sandwich, Analysis, Journal of Applied Mechanics 1971

19. D. Krajcinovic, Sandwich Beams with Arbitrary Boundary ConditionsJournal of Applied Mechanics 1975873 EOF

20. DiTaranto RA.Static Analysis of a Laminated BeamJournal of Engineering for Industry1973755 EOF

21. Johnson AF, Sims GD.Mechanical Properties and Design of Sandwich Materials. Composites1986

22. R. Teti, G. Caprino, First International Conference on Sandwich Construction1989conference proceeding. Mechanical Behavior of Structural Sandwiches. Stockholm, Sweden, 5367

23. K. Lingaiah, B. G. Suryanarayana, Strength and Stiffness of Sandwich Beams in BendingExperimental Mechanics1 EOF7 EOF

24. N. Gupta, E. Woldesenbet, Microscopic Studies of Syntactic Foams Tested Under Three-Point Bending Conditions. American Society of Mechanical Engineers 20021147152

25. R. Maharsia, N. Gupta, H. D. Jerro, Investigation of flexural strength properties of rubber and nanoclay reinforced hybrid syntactic foamsMaterials Science and Engineering: A 2006249 EOF

26. Ravi. Kishore, S. Sankaran, Short, Short-Beam Three-Point Bend Test Study in Syntactic Foam. Part III: Effects of Interface Modification on Strength and Fractographic FeaturesJournal of Applied Polymer Science2005687 EOF

27. Avila AF.Failure mode investigation of sandwich beams with functionally graded coreComposite Structures2007323 EOF

28. C. S. Karthikeyan, S. Sankaran, Kishore, Influence of chopped strand fibres on the flexural behavior of a syntactic foam core system. Polymer International 2000

29. C. S. Karthikeyan, S. Chedarampet, S. Sankaran, Kishore, Flexural Behaviour of Fibre-Reinforced Syntactic FoamsMacromolecular Materials and Engineering200560 EOF

30. C. S. Karthikeyan, S. Sankaran, Kishore, Investigation of bending modulus of fiber-reinforced syntactic foams for sandwich and structural applicationsPolymers for Advanced Technologies2007254 EOF256 EOF

31. N. Gupta, E. Woldesenbet, Characterization of Flexural Properties of Syntactic Foam Core Sandwich Composites and Effect of Density VariationJournal of Composite Materials20052197 EOF2212 EOF

32. Montgomery DC.Design and analysis of experimentsNew York: Wiley; 2001

33. T. H. Ferrigno, H. S. Katz, J. V. Milewski, editor, Handbook of fillers and reinforcements for plastics. New York:Van Nostrand Reinhold; 197819786671

34. S. M. Kulkarni, D. Anuradha, C. R. L. Murthy, Kishore, Analysis of filler-fibre interaction in fly ash filled short fibre-epoxy

composites using ultrasonic NDEBulletin of Materials Science2002137 EOF140 EOF

35. Pedlow JW., Torrey S., editor. Cenospheres, in Coal ash utilization Fly ash, Bottom ash and Slag.New Jersy:Noyes; 19781978353362

36. R. Mohapatra, Rao. Rajagopala, Some aspects of characterisation, utilisation and environmental effects of fly ashJournal of Chemical Technology and Biotechnology20019 EOF26 EOF

37. Kulkarni. S. M. Kishore, S. Sharathchandra, D. Sunil, On the use of an instrumented set-up to characterize the impact behavior of an epoxy system containing varying fly ash content.Polymer testing 2002

38. R. Mohan, Shridhar. M. K. Kishore, R. M. V. G. K. Rao, Compressive strength of jute-glass hybrid fibre compositesJournal of Materials Science Letters198299 EOF102 EOF

39. A. N. Shah, S. C. Lakkad, Mechanical properties of jute-reinforced plasticsFibre Science and Technology198141 EOF46 EOF

40. Gowda TM, Naidu ACB, Rajput Chhaya.Some mechanical properties of untreated jute fabric reinforced polyester compositesComposites Part AApplied Science and Manufacturing 1999277 EOF284 EOF

41. Rengarajan. Mohan, Jute. Kishore-Glass, composites. sandwich, Journal of Reinforced plastics and composites 1985

42. ASTM C393.Standard Test Method for Flexural Properties of Flat Sandwich Construction. ASTM International, PA, USA, 2000

43. N. Gupta, Kumar. Sandeep, Mueller. B. J. Gupta, Analysis of a functionally graded particulate composite under flexural loading conditions.Materials Science and Engineering: A 2008

44. ASTM D3039.Standard Test Method for Tensile properties of polymer matrix composite materials.ASTM International, PA, USA, 2008

45. ASTM D792.Standard test methods for density and specific gravity (Relative density) of plastics by displacement.ASTM International, PA, USA, 2008

3

CHAPTER

FAILURE TEST AND FINITE ELEMENT SIMULATION OF A LARGE WIND TURBINE COMPOSITE BLADE UNDER STATIC LOADING

Xiao Chen [1,2,3,*], Wei Zhao [3], Xiao Lu Zhao [1,2,3] and Jian Zhong Xu [1,2,3]

[1] *Institute of Engineering Thermophysics, Chinese Academy of Sciences, No.11 Beisihuan West Road, Beijing 100190, China*

[2] *National Laboratory of Wind Turbine Blade Research & Development Center, No.11 Beisihuan West Road, Beijing 100190, China*

[3] *Engineering Research Center on Wind Turbine Blades of Hebei Province, No.2011 Xiangyang North Street, Baoding 071051, Hebei, China*

ABSTRACT

This study presented a failure analysis of a 52.3 m composite wind turbine blade under static loading. Complex failure characteristics exhibited at the transition region of the blade were thoroughly examined and typical failure modes were indentified. In order to predict multiple failure modes observed in the tests and gain more insights into the failure mechanisms of the blade, a Finite Element (FE) simulation was performed using a global-local modeling approach and Progressive Failure Analysis (PFA) techniques which took into account material failure and property degradation. Failure process and failure characteristics of the transition region were satisfactorily reproduced in the simulation, and it was found that accumulated delamination in spar cap and shear web failure at the transition region were the main reasons for the blade to collapse. Local buckling played an important role in the failure process by increasing local out-of-plane deformation, while the Brazier effect was found not to be responsible for the blade failure.

Keywords: wind energy; blade failure; composite; delamination; local buckling; progressive failure

1. INTRODUCTION

With the expansion of wind energy in recent years, the sizes of wind turbine blades have become increasingly large in order to capture more power from wind and further reduce the cost of energy [1,2]. When blades are small tip deflection is a major driver in structural design and blade failure is of less concern. However, as blade sizes grow, the types of failure change and three-dimensional stresses become important [3]. In order to investigate the failure behavior of large blades installed on multi-megawatt (MW) wind turbines, some researchers have carried out failure tests on full-scale blades. Among them, Jorgensen, et al. [4] performed a failure test on a 25 m blade under flap-wise bending emphasizing the geometrical nonlinearity at large deflections. Jensen et al. [5–7] tested a 34 m blade and its load-carrying box girder under flap-wise bending and found that the

Brazier effect induced a large deformation in the spar cap and the further delamination buckling were the cause leading to structural collapse. Overgarrd et al. [8,9] tested a 25 m blade to failure under flap-wise bending and concluded that the ultimate strength of the blade was governed by instability phenomena in the form of delamination and buckling instead of the Brazier effect. Yang et al. [10] studied the structural collapse of a 40 m blade under flap-wise bending and found that debonding of aerodynamic shells from adhesive joints was a main reason for the blade to collapse. Chou et al. [11] investigated a typhoon-damaged composite blade with a blade length close to 39.5 m and showed that the blade failed by delamination and laminate cracking. It can be seen that failure causes varied among blades and little consensus has been reached.

Although several failure incidents of 50+ m blades have been reported in recent years [12], no experimental study focused on the failure of blades with lengths longer than 40 m has been publically reported so far according to the authors' knowledge. It is considered that larger blades may exhibit more complex failure behavior which has neither been observed from the existing studies nor been paid enough attention to in current blade design. Therefore, there is an urgent need of experimental studies on large blades to gain more understanding on their failure behavior. Furthermore, numerical models capable of predicting blade failure in a large structural scale are also essential to complement experimental studies due to the expensive cost associated with full-scale blade tests. Although some researchers [13–15] have successfully investigated failure behavior of composite blades using the Finite Element (FE) method, they primarily focused on in-plane failure mode, i.e., laminate fracture (LF), for which failure criteria based on a two-dimensional stress field are readily available in most commercial FE programs. As observed from the existing failure tests on large composite blades, through-thickness failure mode in the form of delamination appears to be more crucial than the in-plane one. Only a few works, i.e., [8,9], have investigated delamination of blades numerically in a structural scale and the FE program they used was developed in-house. It is therefore of significance to develop a modeling method based on commonly used FE programs with which blade designers are familiar for predicting the failure behavior of large blades.

In order to partially meet the need of experimental and numerical studies on the failure of large blades, this study presents a full-scale failure test of a 52.3 m blade with particular focus on its complex failure characteristics which were associated with both in-plane and through-thickness failure modes and have not been investigated by other studies focusing on smaller blades. In order to predict the complex failure characteristics of the blade, this study proposes a FE modeling method based on a commercial FE program and further investigated the mechanisms leading to the blade failure. It was expected that more understanding on the failure of large wind turbine blades could be obtained from this study.

The contents of this paper are organized as follows: general information and experimental procedures of the blade were described and then failure characteristics were investigated in Section 2; in Section 3 a FE modeling method which was capable of predicting complex failure characteristics of the blade was proposed taking advantage of Progressive Failure Analysis (PFA) techniques and a global-local modeling approach; after assessing numerical models by comparing predicted structural response with available experimental measurements, failure process and failure characteristics of the blade were reproduced in Section 4; based on failure test and numerical simulation the root cause of failure and the Brazier effect were examined to understand the failure mechanisms of the blade in Section 5; finally, the findings obtained from this study are summarized in Section 6.

2. FULL-SCALE FAILURE TEST

Commercial wind turbine blades have to pass full-scale tests required by blade certification bodies [16,17] in order to start a series production of those blade types. Static bending loads are applied in the test to simulate extreme wind loads the blades are expected to sustain in their design lives. This section presents general information and the experimental procedure of the blade under concern and investigated its complex failure characteristics through post-mortem observations.

2.1. GENERAL INFORMATION OF THE BLADE

The blade was designed for 2.5 MW wind turbines and had a total length of 52.3 m. It was made of glass fabrics and vacuum infused with epoxy resin. Typical construction of the blade cross-sections is shown in Figure 1. Spar caps were made of composite laminates and designed to carry bending moments, while leading edge (LE) panel and aft panel were made of sandwich constructions with PVC foam cores and designed to provide an aerodynamic airfoil shape. Shear webs were also sandwich constructions and designed to support two spar caps and transfer shear forces.

Throughout this paper span-wise locations of the blade were normalized by the total length of the blade and denoted as "Lxxxx". For example, a span-wise location of 4 m was denoted as "L0.076" in Figure 2, which shows composite layups at representative blade locations. The blade surfaces contained two types of triaxial laminates denoted as "Triaxial_7" and "Triaxial_1"; spar caps contained a large amount of unidirectional laminates and shear webs contained biaxial laminates as sandwich skins.

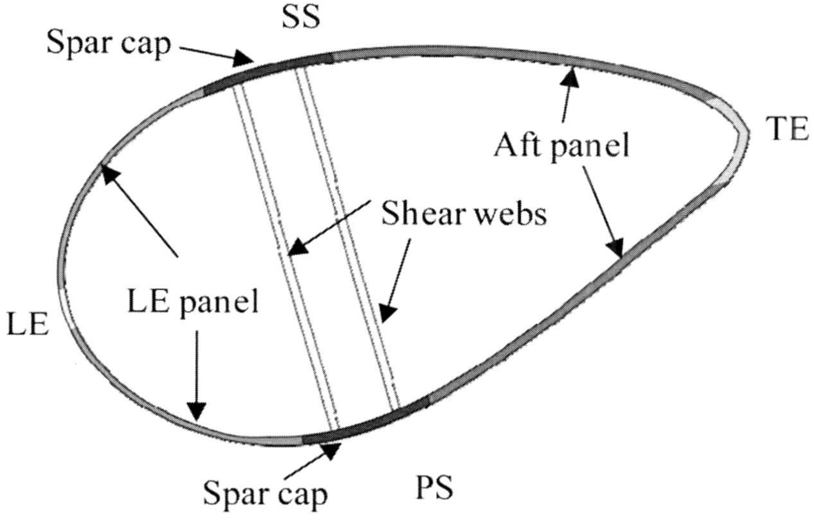

Figure 1. Typical construction of blade cross-section.

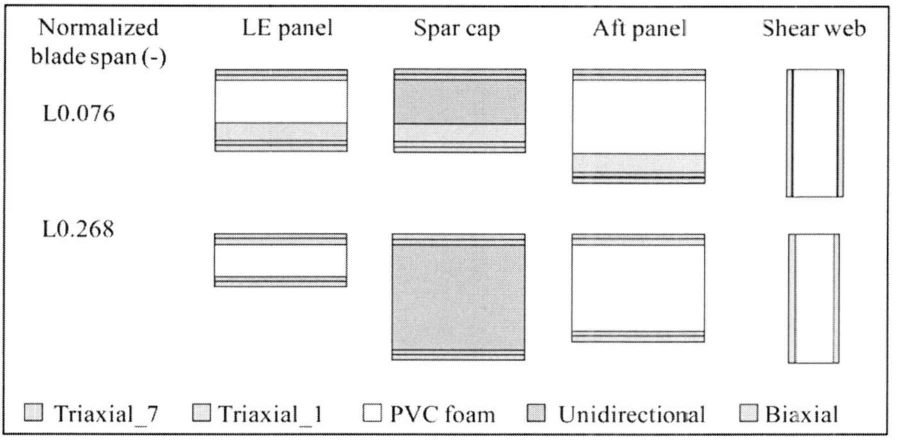

Figure 2. Representative composite layups of the blade.

2.2. Experimental Procedure

In general, two sets of static bending tests were performed on this blade, the first set was based on design loads for 2.5 MW wind turbines, and it followed the test requirements for full-scale wind turbine blades according to certification bodies. The second set was based on design loads for 3.0 MW wind turbines and it was performed with the intention to overload the blade and examine possible failure. All test cases and their sequence are shown in Table 1. Loading directions are schematically shown in Figure 3 with typical cross-sectional profiles of the blade superposed. It is noted that the test sequence of the blade was determined to comply with an order of counterclockwise rotation of the blade axis. The blade was root-fixed on a test stand as a cantilever beam and cranes applied upward pulling forces as shown in Figure 4 for the test case "F_Max_3.0", after one test case was finished, the blade was rotated along its longitudinal axis to a prescribed position for a subsequent test case.

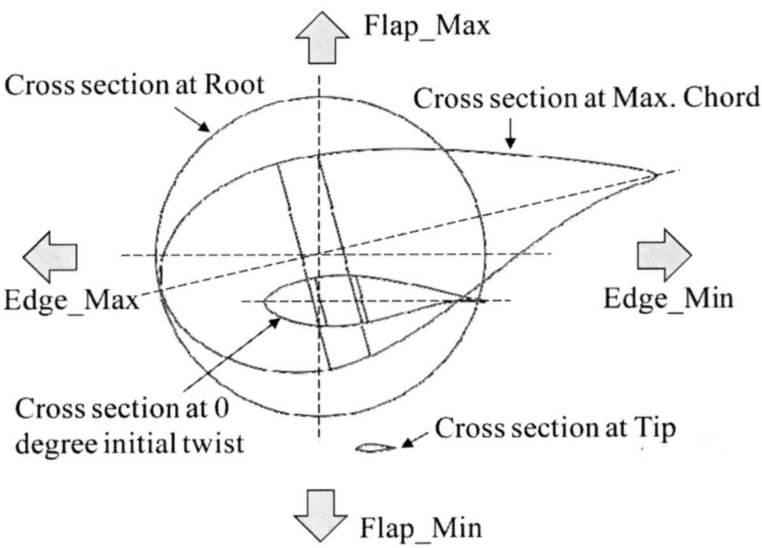

Figure 3. Loading direction of test cases.

Figure 4. Test scene of the blade in the test case "F_Max_3.0".

Table 1. Load test history of the blade.

Test sequence	Test case	Turbine Class	Loading direction
1	F_Min_2.5	2.5 MW	Flap_Min
2	E_Min_2.5		Edge_Min
3	F_Max_2.5		Flap_Max
4	E_Max_2.5		Edge_Max
5	F_Min_3.0	3.0 MW	Flap_Min
6	E_Min_3.0		Edge_Min
7	F_Max_3.0		Flap_Max
8 *	E_Max_3.0		Edge_Max

* Test case "E_Max_3.0" was not tested due to the final blade failure in "F_Max_3.0".

Three cranes and four cranes were used for 2.5 MW loading and 3.0 MW loading, respectively. Rubber pads with approximately 200 mm width and 15 mm thickness were installed between blade surfaces and loading saddles to reduce stress concentration at loading locations. The target test bending moments applied to the blade were normalized by the maximum root moment in the test case "F_Max_2.5" and they are shown in Figure 5. It can be seen that the target test loads of "F_Max_2.5" were basically same as those of "F_Max_3.0". Test loads were applied in a stepwise form following a loading procedure of 0%, 40%, 60%, 80%, 100% and then an unloading procedure of 80%, 60%, 40%, 0% of the target loads in each test case. There was no communication among cranes which applied pulling force simultaneously to obtain each prescribed load level and then held for ten seconds as required by [16,17]. During the test, applied loads were recorded by load cells mounted on the cranes, deflections of the blade were measured at loading saddle locations and at blade tip using draw-wire displacement transducers, longitudinal strains were recorded by strain gauges located along the middle axis of spar caps. Average values of measurements during ten seconds were used to represent the structural response of the blade at the prescribed load levels.

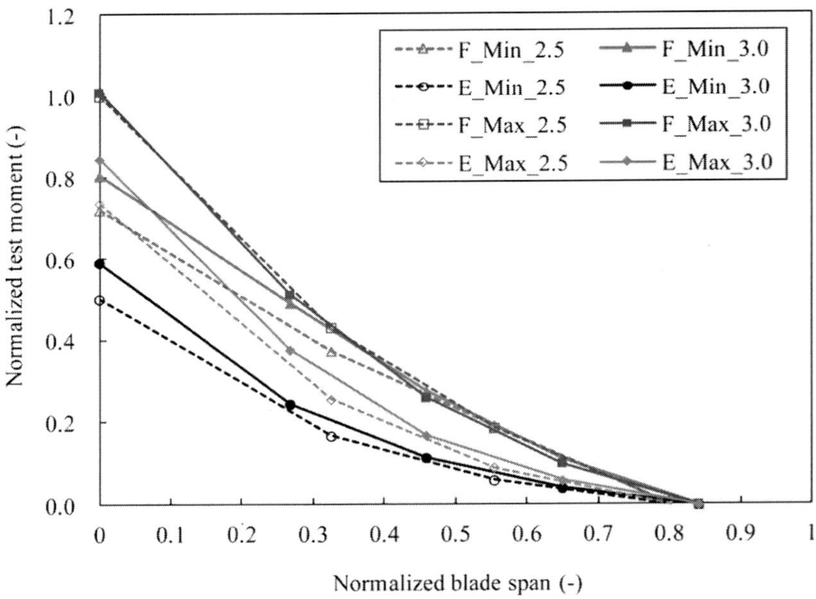

Figure 5. Applied target bending moments.

2.3. Post-Mortem Observations and Discussion

After the successful completion of the 2.5 MW loading set for blade certification, the 3.0 MW loading set was followed. At 100% of the target test loads in the test case "E_Min_3.0", trailing edge (TE) and the adjacent aft panels failed at a transition region where the cross-sectional geometry of the blade transited from a circular shape at the blade root to an airfoil shape at the maximum chord. After unloading the test case "F_Max_3.0" was followed and the blade collapsed due to a final failure at the transition region at around 90% of the target test loads. Because the target test loads of "F_Max_2.5" and "F_Max_3.0" were basically the same, the final failure load of the blade was actually 10% smaller than the load once carried by the blade.

Thorough investigation was then performed on the entire blade. It was found that major failure regions were located from L0.067 to L0.105 as shown in Figure 6a. Although failed regions exhibited a complex form of failure, some typical failure modes, i.e., laminate fracture (LF), composite delamination (DL), and sandwich skin-core debonding (DB) could be identified as shown in Figure 6b,c where visually apparent debonding fronts were depicted with dashed curves, and intersections between spar cap and sandwich panels were depicted with dashed lines. A close-up of the suction side (SS) of the blade at L0.076 shows local failure characteristics in Figure 6d, and it can be seen that LF occurred at the intersection between the spar cap and aft panel, and skin laminates of spar cap were subjected to LF as well as DL.

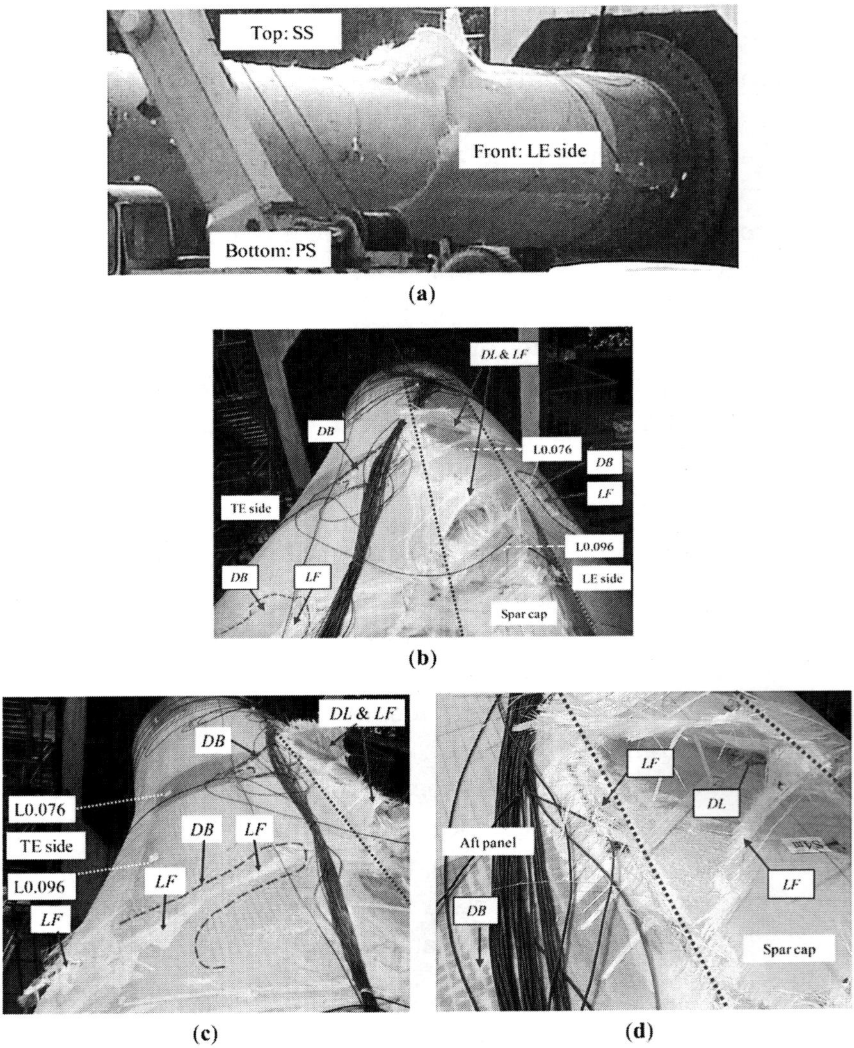

Figure 6. External failure of the blade. (**a**) Final failure of the blade at the transition region; (**b**) View#1 Typical failure modes at SS; (**c**) View#2 Typical failure modes at SS; (**d**) Local failure features at L0.076.

Inspection on the blade interior was also made, and it was found that the rear shear web at the location from L0.076 to L0.096 was completely failed with an approximate fracture angle of 45° to the blade axis, see Figure 7a. TE at L0.092 exhibited considerable failure with combined failure modes of LF, and DB as shown in Figure 7b.

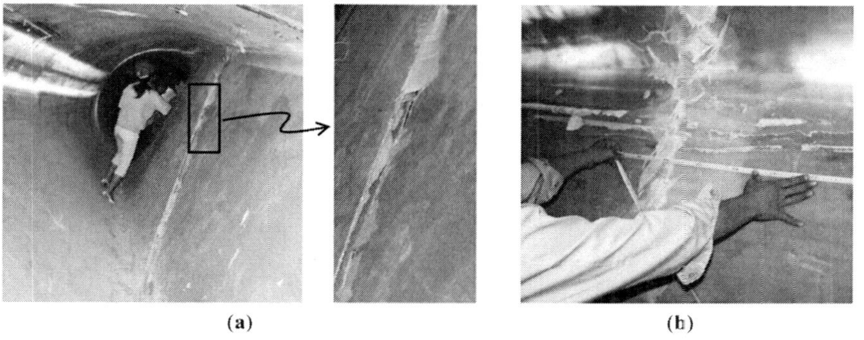

Figure 7. Internal failure of the blade. (**a**) Rear shear web failure; (**b**) TE failure.

Furthermore, the blade was sectioned at L0.076 to facilitate examination on its cross-section. It was found that DL of spar cap occurred not only at triaxial laminates constituting blade skins but also at unidirectional laminates as shown in Figure 8a. Investigation on aft and LE panels revealed DB and associated core shear failure (CF) at sandwich constructions and the failure surface of DB propagated at a region roughly 1 to 2 mm underneath the interface between skin laminates and core materials as shown in Figure 8b.

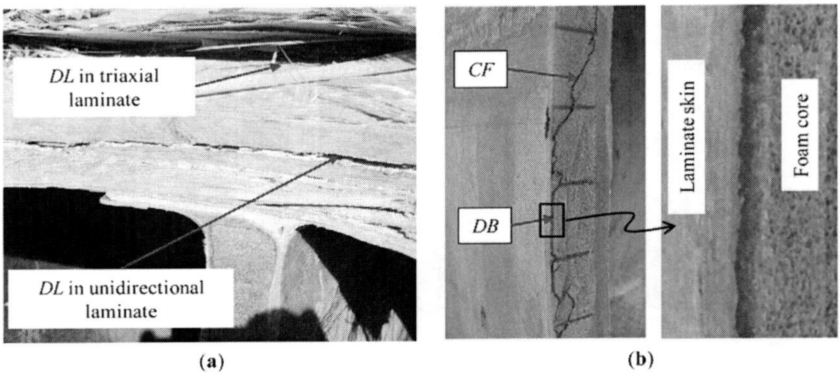

Figure 8. Failure observed at L0.076 cross section. (**a**) The SS spar cap; (**b**) LE panel.

From post-mortem observations, it was obvious that the transition region was pervaded by multiple failure modes. Among them DL and DB were essentially related to mechanical properties and stress states of interfaces between constituent layers and they could be categorized as interfacial (or through-thickness) failure. Another failure mode LF although commonly considered to be a form of in-plane failure was found at the intersection where geometric and material discontinuities induced interlaminar stresses inevitably affected the characteristics of LF. Because both interfacial and interlaminar stresses are closely associated with through-thickness stresses, it can be concluded that through-thickness stresses played an important role in the complex failure characteristics with multiple failure modes of the blade.

Furthermore, it is noted that spar caps with a large amount of unidirectional laminates governed the bending stiffness and the overall strength of the blade, and shear webs supported two halves of aerodynamic shells and transferred shear forces in the blade. From post-mortem observations, it can be seen that among various failure modes exhibited at the transition region delamination of unidirectional laminates in the SS spar cap and the failure of shear web were most responsible for the complete loss of the load-carrying capacity of the blade.

3. FE MODELING METHOD

Although post-mortem observations provided some understanding of the failure characteristics of the blade, the failure process and failure mechanisms have not been clarified yet. In order to gain more insights into the failure of the blade, a numerical model is necessary to complement the experimental study. This section, aimed at developing a FE modeling method for numerical study, first presents the considerations that have to be taken into account in the simulation and then describes material models used in the PFA techniques and a global-local approach for blade modeling. Furthermore, numerical implementation of the simulation in a commercial FE program is also presented.

3.1. Considerations in FE Modeling

Based on the failure observations from the test, it was evident that two features have to be incorporated in FE modeling in order to accurately simulate the failure process and failure characteristics of the blade. The one feature is that the FE model to be constructed should consider the effect of loading history on structural performance as the final failure load was found to be smaller than the load once carried by the blade, implying the adverse effect of loading history on its ultimate load-carrying capacity. This feature can be achieved by using well-established PFA techniques [18] which consist of two parts, i.e., failure criteria indicating occurrence of material failure and material property degradation rules representing residual properties of material once failure criteria are satisfied. Although the PFA techniques have been used in many works [19,20] to investigate failure processes of composite materials and structural components, they were rarely applied to analyze large composite blades on a structural scale. In the current study the PFA techniques were used to simulate the loading history of the blade with material failure and property degradation.

The other feature that needs to be included in the FE model are the three-dimensional stresses which enable analysis of complex failure characteristics associated with both in-plane and through-thickness stresses, and this requires solid elements to be used in the model. However, modeling and analyzing an entire blade structure using solid elements is extremely difficult due to the large differences in size scales of the component details and the entire blade. Considering that the blade failure occurred primarily at the transition region, only a local part of the blade had to be modeled and analyzed using solid elements. Although the local solid element model appeared to be a promising solution to reduce modeling difficulty, the actual loads applied to the entire blade instead of the transition region were explicitly known. In order to obtain loads sustained by the transition region, it was decided to first construct a global shell element model of the entire blade representing its load status in the test and back-calculate loads sustained by the transition region, and then these loads were applied to the local solid element model and detailed failure analysis was subsequently performed.

3.2. Material Models

3.2.1. Material Properties

Experimental results of in-plane properties from material tests are shown in Table 2. Due to commercial concerns from blade designer, all modulus and strength parameters of materials were normalized by the corresponding longitudinal properties of each material. For the local solid element model, through-thickness properties were needed to perform the analysis. However, material properties in the through-thickness (or "3") direction were not tested in this study, so some assumptions were used to estimate these values as shown in Table 3 where the properties of unidirectional composites were regarded to be proportional to their in-plane ones, and the proportional factors were determined according to [21], in which a similar glass/epoxy unidirectional thick laminate has been systematically tested. The strength properties of two types of triaxial composites in the "3" direction were assumed to be the same as those of unidirectional composites considering that there was no fiber reinforcement aligned in this direction and strength properties were largely controlled by the matrix in composites.

Table 2. Material in-plane properties used in both the global shell and local solid element models.

Name	Elastic constants				Strength parameters				
	E_{11}	E_{22}	G_{12}	μ_{12}	σ_{11t}^u	σ_{11c}^u	σ_{22t}^u	σ_{22c}^u	σ_{12}^u
Unidirectional	1	0.33	0.09	0.21	1	−0.59	0.04	−0.20	0.06
Triaxial_7	1	0.73	0.51	0.47	1	−0.94	0.31	−0.86	0.43
Triaxial_1	1	0.47	0.30	0.36	1	−0.78	0.33	−0.21	0.15
Biaxial	1	1.00	0.82	0.49	1	−1.00	1.00	−1.00	1.51
PVC foam	1	1.00	0.63	0.30	1	−0.50	1.00	−0.50	0.63

Where subscripts 11 and 22 mean longitudinal and transverse direction, respectively; subscript 12 means in-plane shear direction; subscripts t and c mean tensile and compressive direction, respectively; superscript u indicates ultimate strength parameters.

Table 3. Material through-thickness properties used in the local solid element model.

Name	Elastic constants					Strength parameters			
	E_{33}	G_{13}	G_{23}	μ_{13}	μ_{23}	σ_{33c}^u	σ_{33t}^u	σ_{13}^u	σ_{23}^u
Unidirectional	E_{22}	$1.08G_{12}$	G_{12}	μ_{12}	$1.33\mu_{12}$	$1.03\sigma_{22c}^u$	$0.71\sigma_{22t}^u$	$0.97\sigma_{12}^u$	$0.82\sigma_{12}^u$
Triaxial_7	E_{22}	G_{12}	G_{12}	μ_{12}	μ_{12}	=	=	=	=
Triaxial_1	E_{22}	G_{12}	G_{12}	μ_{12}	μ_{12}	=	=	=	=
Biaxial	N.A.	N.A.	N.A.	N.A.	N.A.	N.A.	N.A.	N.A.	N.A.
PVC foam	E_{22}	G_{12}	G_{12}	μ_{12}	μ_{12}	σ_{22c}^u	σ_{22t}^u	σ_{12}^u	σ_{12}^u

Note: "=" means that the value is the same as that for unidirectional composite; "N.A." means "not applicable", it is because biaxial materials are used in shear webs, which are modeled with shell elements, and their through-thickness properties are not needed in the FE model.

In order to predict interfacial failure, i.e., DL and DB, in the solid element model, material properties of interfaces need to be known. In this study, all elastic constants and strength parameters of interfaces between two composite lamina layers were assumed to be equal to the average values of material properties of two lamina layers. As for interfaces between composite lamina and PVC foam core, material properties were assumed to be the same as PVC foam considering that the failure surface of skin-core debonding occurred in PVC foam as observed in the test.

3.2.2. Failure Criteria

Three typical failure modes, i.e., LF, DL, and DB, observed in the failed blade were considered in FE modeling. For LF, the Tsai-Wu failure criterion [22] in a three-dimensional stress field, as expressed in Equation (1), was employed in this study. The criterion is widely used to predict LF and has several advantages. It allows for interaction among stress components (analogous to the von Mises criterion for isotropic materials), it takes into account differences between tensile and compressive strength, and it is operationally simple and readily amenable to computational procedures [23]:

$$F.I._{3d} = F_1\sigma_{11} + F_2\sigma_{22} + F_3\sigma_{33} + 2F_{12}\sigma_{11}\sigma_{22} + 2F_{13}\sigma_{11}\sigma_{33} + 2F_{23}\sigma_{22}\sigma_{33} + F_{11}\sigma_{11}^2 + F_{22}\sigma_{22}^2 + F_{33}\sigma_{33}^2 + F_{44}\sigma_{23}^2 + F_{55}\sigma_{13}^2 + F_{66}\sigma_{12}^2 \qquad (1)$$

where,

$F_1 = 1/(\sigma_{11t}^{u} - \sigma_{11c}^{u})$, $F_2 = 1/(\sigma_{22t}^{u} - \sigma_{22c}^{u})$, $F_3 = 1/(\sigma_{33t}^{u} - \sigma_{33c}^{u})$, $F_{12} = -0.5/(\sigma_{11t}^{u}\sigma_{11c}^{u}\sigma_{22t}^{u}\sigma_{22c}^{u})^{0.5}$, $F_{13} = -0.5/(\sigma_{11t}^{u}\sigma_{11c}^{u}\sigma_{33t}^{u}\sigma_{33c}^{u})^{0.5}$, $F_{23} = -0.5/(\sigma_{22t}^{u}\sigma_{22c}^{u}\sigma_{33t}^{u}\sigma_{33c}^{u})^{0.5}$, $F_{11} = 1/(\sigma_{11t}^{u}\sigma_{11c}^{u})$, $F_{22} = 1/(\sigma_{22t}^{u}\sigma_{22c}^{u})$, $F_{33} = 1/(\sigma_{33t}^{u}\sigma_{33c}^{u})$, $F_{44} = 1/(\sigma_{23}^{u})^2$, $F_{55} = 1/(\sigma_{13}^{u})^2$, $F_{66} = 1/(\sigma_{12}^{u})^2$; and $F.I._{3d}$ was defined as failure index in a three-dimensional stress field, and a value larger than 1 indicates a composite failure; σ_{11}, σ_{22}, and σ_{33} are normal stresses of composites in longitudinal, transverse and thickness direction, respectively; σ_{12}, σ_{23}, and σ_{13} are shear stresses in in-plane, transverse, and axial direction, respectively. The superscript u means ultimate strength of the material, and subscripts t and c mean tensile and compressive direction, respectively.

This failure criterion is mode-independent as it does not directly identify the nature of material damage. In order to apply appropriate material degradation rules, the nature of LF needs to be differentiated. According to Christensen [24], the term $F_1\sigma_{11} + F_{11}\sigma_{11}^2$ indicates a fiber-controlled failure and it was employed to identify the fiber (or matrix)-controlled failure mode if it is larger (or smaller) than the summation of other terms in Equation (1) when failure indices exceed 1.

DL and DB were regarded as two different failure modes emphasizing their locations of occurrence, whereas they shared a similar failure feature that two constituent layers separated and the properties of their interface dominated failure behavior. Therefore, a same quadratic failure criterion based on Ye [25] as expressed in Equation (2) was employed to predict the occurrence of DL and DB:

$$F.I._{\text{iff}} = (\sigma_{33}{}^{+}/\sigma_{33\text{t}}{}^{\text{u}})^2 + (\sigma_{13}/\sigma_{13}{}^{\text{u}})^2 + (\sigma_{23}/\sigma_{23}{}^{\text{u}})^2 \quad (2)$$

where, $F.I._{\text{iff}}$ was defined as a failure index of interfacial failure, and a value larger than 1 indicates an interfacial failure, $\sigma_{33}{}^{+}$, σ_{13}, and σ_{23} are normal tensile stress, axial shear stress, and transverse shear stress, respectively; $\sigma_{33t}{}^{u}$, $\sigma_{13}{}^{u}$, and $\sigma_{23}{}^{u}$ are the corresponding ultimate strength. It is noted that all stress variables and strength parameters refer to the interface of two constituent layers.

3.2.3. Material Degradation Rules

Once failure criteria were satisfied, material properties were degraded according to certain rules representing the effect of material damage.

Some material degradation rules in the form of multiplying factors [26–28] as shown in Table 4 have been used in the failure analysis of composites, and these multiplying factors were applied to the intact material properties to obtain the degraded ones once material failure criteria were met. In the present study, the degradation rule adopted by Apalak et al. [27] was used with changing G_{13} value for the matrix-controlled LF from 0 to 1 regarding insignificant effect of matrix failure on transverse normal shear modulus. The value of zero was replaced by a very small value in order to avoid numerical errors in the FE analysis. The multiplying operation was only performed once at the first occurrence of the corresponding failure mode and the properties of materials hereafter remained constant, representing the irreversible nature of material damage.

Table 4. Material degradation rules used for PFA.

Failure mode	Model	E_{11}	E_{22}	E_{33}	G_{12}	G_{13}	G_{23}	μ_{12}	μ_{13}	μ_{23}
LF (Fiber-controlled)	Nagesh [26]	0.1	0.1	0.1	0.1	0.1	0.1	0	1	1
	Z.G. Apalak *et al.* [27]	0	0	0	0	0	0	0	0	0
	K.I. Tserpes *et al.* [28]	0	0	0	0	0	0	0	0	0
	Present	0	0	0	0	0	0	0	0	0
LF (Matrix-controlled)	Nagesh [26]	1	0.1	0.1	1	1	0.1	0	0	1
	Z.G. Apalak *et al.* [27]	1	0	0	1	0	0	0	0	1
	K.I. Tserpes *et al.* [28]	1	0	1	1	1	1	0	1	1
	Present	1	0	0	1	1	0	0	0	1
DL	Nagesh [26]	1	0.1	0.1	1	1	0.1	0	0	1
	Z.G. Apalak *et al.* [27]	1	1	0	1	0	0	1	0	0
	K.I. Tserpes *et al.* [28]	1	1	0	1	0	0	1	0	0
DL&DB	Present	1	1	0	1	0	0	1	0	0

3.3. FE Models

3.3.1. The Global Shell Element Model

The global shell element model was constructed in order to obtain loads sustained by the transition region by simulating a full-scale blade in the test. In this study, the general FE program Abaqus [29] which is commercially available was used. In total, about 47,300 "S4R" shell elements with a typical mesh size of 100 mm × 100 mm were used in the model. "S4R" is a 4-node, quadrilateral, stress/displacement shell element with reduced integration and a large-strain formulation. Before the final mesh size was determined, a mesh convergence study has been conducted. It was found that when the blade was modeled with typical mesh sizes of 150 mm × 150 mm, 100 mm × 100 mm, and 50 mm × 50 mm, the results of the first natural frequency between 100 mm × 100 mm mesh and 50 mm × 50 mm mesh were below 1%, and the results of the first bucking eigenvalue between two meshes were below 2%, therefore, a mesh size of 100 mm × 100 mm was deemed sufficient.

Layered orthotropic materials were applied to the blade model according to the layup scheme for blade manufacturing. A fixed boundary was applied at the blade root and point loads with resultants equal to the forces recorded from load cells were equally distributed on the surfaces where spar caps of the blade and rubber pads in the loading saddles are contacted representing actual loading setup used in the test, as shown in Figure 9. Loads were applied incrementally up to the target test loads and geometric nonlinearity was taken into account in the model to capture large deformation. Three reaction forces and three reaction moments at the blade root were extracted in every load step of each test case.

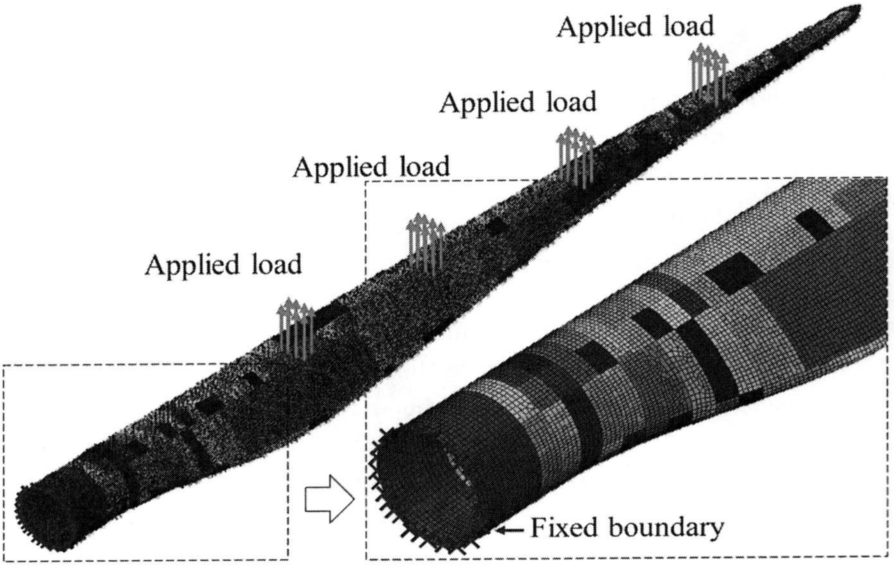

Figure 9. The global shell element model with applied loads in the test case "F_Max_3.0".

3.3.2. The Local Solid Element Model

In the present work, major blade failure of primary interest was located at the transition region approximately from L0.067 to L0.105 of the blade. Therefore, only the inboard part of the blade up to L0.134 was constructed in a local FE model. Both shell elements for shear webs and solid elements for the rest were used in the model. Nevertheless, this FE model was regarded as the local solid element model considering its different capacity to predict blade failure from the global shell element model.

There were about 25,900 solid "C3D8RC3" elements out of the total 33,000 elements used to mesh the local part. "C3D8RC3" is a 8-node, three-dimensional, linear, continuum stress/displacement element with reduced-integration. A typical mesh size was 45 mm × 45 mm for shell elements and 45 mm × 45 mm × t_i for solid elements, and t_i varied according to the local thickness of blade external shells. Only one solid element was used in the through-thickness direction of the blade shells. The root of the local part was fixed and multiple-point constraints (MPC) [29] were applied to a loading section at L0.134 to introduce appropriate loads through a reference point, which was positioned to have same cross-sectional coordinates as the center of the circular root section as shown in Figure 10.

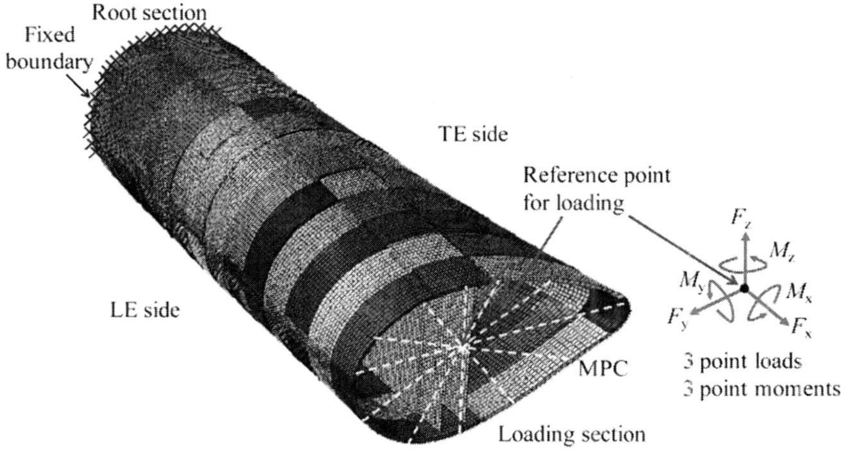

Figure 10. The local solid element model with applied loads back-calculated from shell element model.

Three point loads and three point moments were applied to the reference point in such a way that each reaction force and reaction moment yielded at the blade root was equal to the corresponding ones obtained from the global shell element model which was loaded according to the actual experimental setup. Layered orthotropic materials were assigned to the model. In one solid element, composite laminates were subdivided into several lamina layers with actual ply thicknesses and two types of interface layers with a negligible thickness of 0.001 mm were embedded between lamina layers, i.e., Type 1 interface layer, as well as between composite material and PVC foam core, i.e., Type 2 interface layer, as shown in Figure 11. Interface layers were employed in the solid element model in order to capture interfacial failure modes, i.e., DL and DB, in the FE analysis.

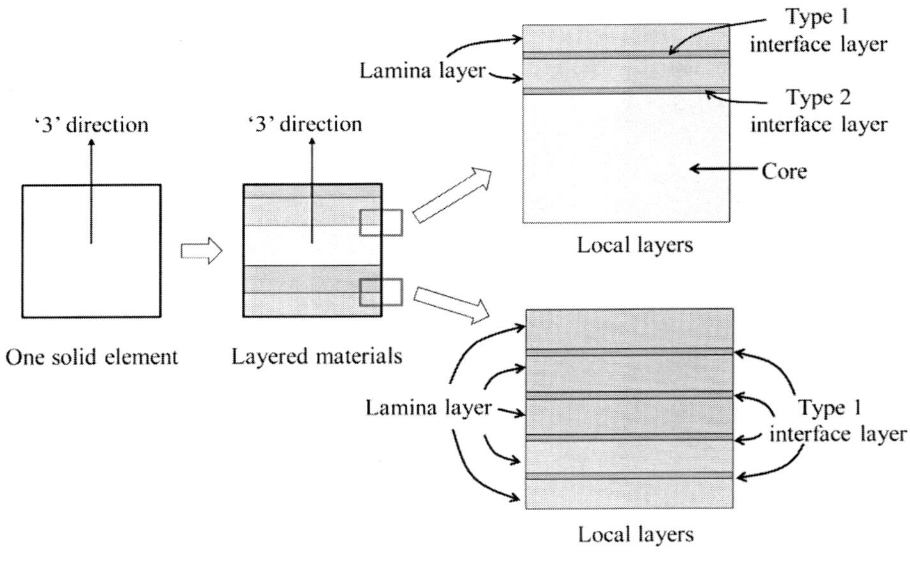

Figure 11. Schematic presentation of layered features in a solid element.

3.4. Numerical Implementation of Simulation

The PFA techniques were implemented in the solid element model to capture the complex failure characteristics observed at the transition region by incorporating user subroutines into Abaqus according to the proposed FE modeling method. The occurrences of material failure were predicted using prescribed failure criteria and material properties upon failure were degraded according to degradation rules in a three-dimensional stress field. All seven test cases were analyzed in a single numerical run following the loading history of the blade. The Riks method was adopted in order to help static equilibrium in the analysis with considerable geometric and material nonlinearities. Material failure indices larger than 1 were updated and recorded in each load increment resembling the progress of material failure in the blade.

4. SIMULATION RESULTS AND DISCUSSION

This section presents results obtained from FE simulation. Two FE models used in the global-local modeling approach were assessed by comparing predicted results with experimental measurements available in the 2.5 MW loading set, and then failure process and the final failure characteristics of the transition region were predicted. Furthermore, taking advantage of the local solid element model, failure mechanisms was investigated and the Brazier effect was clarified in this section.

4.1. Model Assessment

4.1.1. Shell Model Assessment

In order to ensure that the global shell element model was able to represent the loading status of the blade in the test, numerical predictions of blade deflections and strains were compared with experimental measurements in the 2.5 MW loading set. Representatively, deflection measurements of the test case "F_Max_2.5" are shown in Figure 12a, where numerical predictions from the global shell FE model are also plotted. Longitudinal strains along spar caps at both pressure side (PS) and suction side (SS) of the blade are shown in Figure 12b.

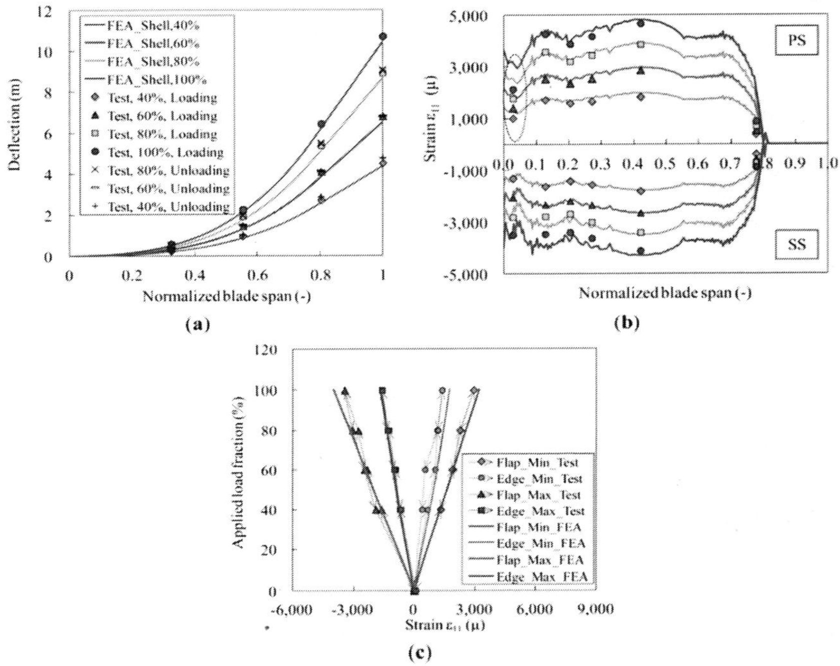

Figure 12. Comparison of structural response of the blade in the 2.5 MW loading set. (**a**) Deflection in the test case "F_Max_2.5"; (**b**) Longitudinal strains in the test case "F_Max_2.5"; (**c**) Longitudinal strains at the SS spar cap at L0.126.

For four test cases in the 2.5 MW loading set, longitudinal strains at SS of the blade at L0.126 are shown in Figure 12c, where arrow signs on dashed lines indicate loading and unloading paths. From these comparisons, it was found that the predicted global deflections and longitudinal strains were in satisfactory agreement with experimental measurements despite relative large difference in strains at one location as indicated with a dashed circle in Figure

12b. Therefore, the global shell element model was considered to be well constructed and can be used to calculate loads sustained by the transition region.

4.1.2. Solid Model Assessment

After applying three point loads and three point moments which back-calculated from the global shell element model to the reference point in the local solid model, the PFA was carried out to reproduce the loading history.

In order to assess the capability of the solid element model to represent the transition region of the blade in the test, the model should have been validated against experimental results, however, due to test data availability at this region, the solid element model was alternatively compared to the shell element model which has been proved to be able to predict deflections and longitudinal strains of the blade with reasonable accuracy. The test case "F_Max_2.5" was considered for the comparison representatively as shown in Figure 13 where available experimental measurements were plotted. It was found that both blade deflections and longitudinal strains obtained from the global shell element model and the local solid element model showed general agreement, despite relative large differences in strain at a few locations. Based on the comparison, it was regarded that the solid element model was loaded appropriately and was able to represent the transition region in the actual test.

4.2. Failure Prediction

4.2.1. Predicted Failure Process

Using the proposed modeling method, progressive failure process of the transition region was predicted. Due to massive results generated in the simulation, only results at the peak load of each test case were presented in this study. The locations with failure indices larger than 1 indicating material failure are shown in Figure 14.

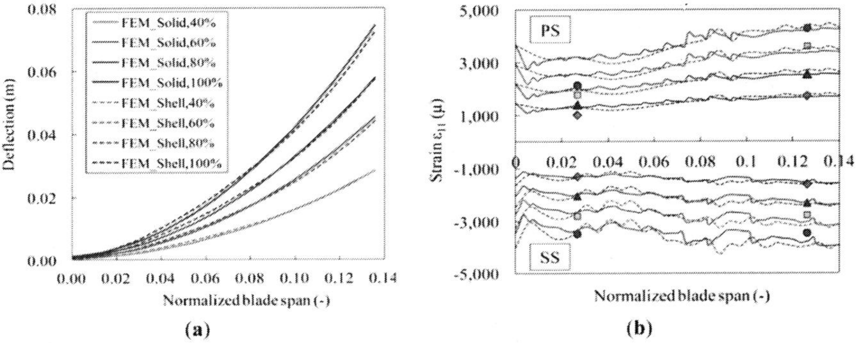

Figure 13. Comparison of results obtained from shell and solid element models. (**a**) Blade deflection; (**b**) Longitudinal strains on spar cap.

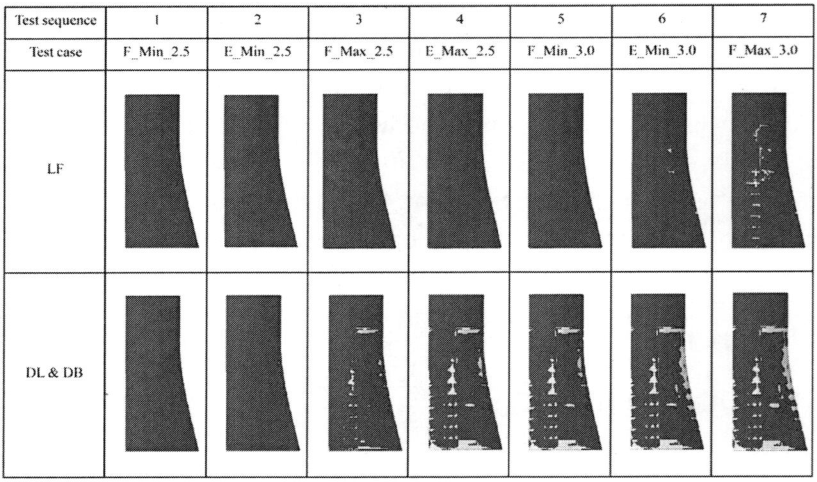

Figure 14. Progressive failure at the peak load of each test case.

It was found that LF was negligible until the test case "E_Min_3.0", whereas interfacial failure in the form of DL and DB has been developed at a preceding test case "F_Max_2.5" when SS of the transition region was primarily subjected to compressive force due to applied flap-wise bending. These interfacial failure continued to progress around TE and spar cap at a subsequent test case "E_Max_2.5". When edge-wise bending was applied in the test case "E_Min_3.0", LF can be distinctly observed at TE and aft panels and interfacial failure spread over the transition region at various locations. Subsequently, when SS of the transition region was once again subjected to compressive force in the test case "F_Max_3.0", LF progressed rapidly and interfacial failure was also further developed.

4.2.2. Predicted Final Failure Characteristics

The predicted final failure of the transition region was presented in a perspective view in Figure 15 for more discussion. It can be seen from Figure 15a that major regions with LF were located at SS of the blade from L0.061 to L0.086, which agreed satisfactorily with the experimental observations. Moreover, some complex failure characteristics found in the test were also well captured by the simulation, such as LF at the intersection between spar cap and aft panel. LF was also predicted to occur at intersecting regions of blade root, spar cap and aft panel although this was not visually detected in the test.

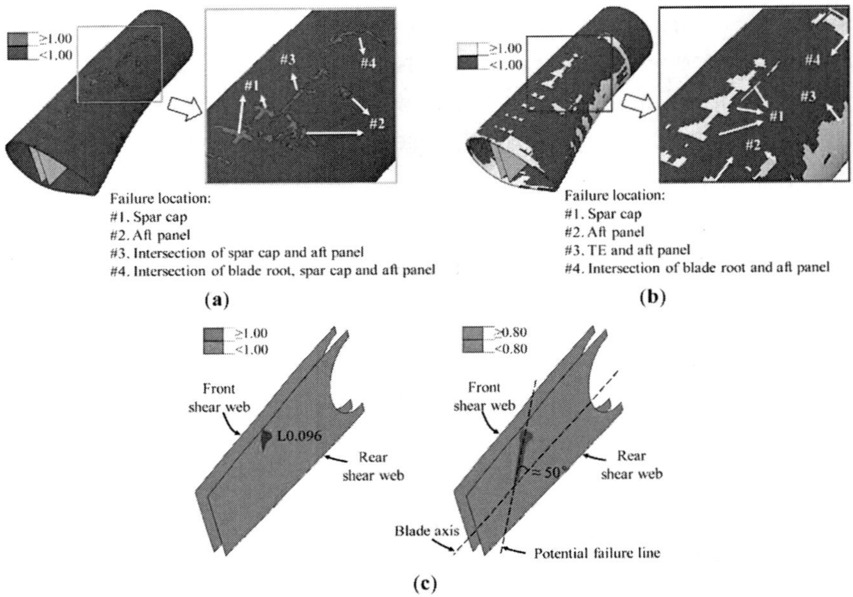

Figure 15. The final failure predicted from FE simulation. (**a**) LF at blade shells; (**b**) DL and DB failure; (**c**) LF at the rear shear web.

For interfacial failure as shown in Figure 15b, significant DL was found at spar cap, TE and the adjacent aft panel where similar failure characteristics were observed in the test although the scale of failure locations appeared to be overestimated. LF of the rear shear web was predicted to initiate at the top of the rear shear web at L0.096 as shown in Figure 15c. The failure region with failure indices larger than 1 was relatively small, but a contour with failure indices larger than 0.8 showed that a potential failure line originated from the initial failure location and inclined approximately 50 degree to the blade axis, which basically agreed with failure characteristics found from the post-mortem observations.

In order to examine the effect of different material degradation rules on the final failure prediction, the present material degradation rule was replaced with that used by Nagesh [26] and one new simulation was performed. The contours of the final failure obtained from two analyses were compared in Figure 16. It can be seen that failure locations predicted from the analysis with the Nagesh rule were relatively larger than those with the present one. For LF the locations with larger failure were spar cap and aft panels, and for interfacial failure, they were TE and the adjacent aft panel. Although there was a difference in the scale of the failure locations, the two analyses predicted similar results in terms of major failure locations and failure characteristics found at the transition region of the blade.

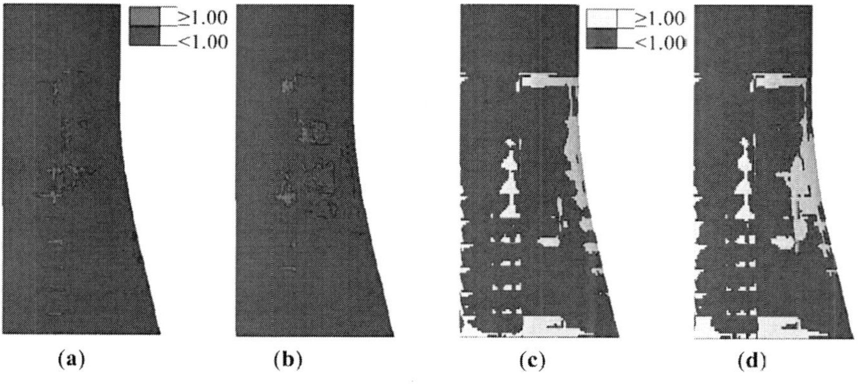

Figure 16. Effect of different material degradation rules on the final failure prediction. (**a**) LF_Present; (**b**) FL_Nagesh [26]; (**c**) DL&DB_Present; (**d**) DL&DB_Nagesh [26].

5. FAILURE MECHANISMS

This section presented investigation on the root cause of blade failure according to experimental observations and numerical simulation. The Brazier effect, which has been argued for its role in the blade failure among several studies [6,8,10] was also examined using the numerical model.

5.1. Root Cause of the Blade Failure

Taking advantage of experimental observations and numerical simulation, the mechanisms leading to the final failure were investigated. In the 2.5 MW loading set, the transition region accumulated material damage in the form of interfacial failure, i.e., DL of unidirectional laminates in spar cap and DB in sandwich constructions. When the 3.0 MW loading set with large load amplitudes in the first two test cases, i.e., "F_Min_3.0" and "E_Min_3.0", were applied to the blade, interfacial failure was further developed in spar cap and TE, and LF also occurred at aft panel and TE. At the test case "F_Max_3.0", in which SS of the blade was once again subjected to compressive force in flap-wise bending, the previously developed LF continued to progress around the spar cap region with significant interfacial failure resulting in a drastic decrease of the bending stiffness and the overall strength. Meanwhile, the rear shear web started to fail further reducing its capacity to transfer shear forces. As a consequence, the blade lost its load-carrying capacity at the transition region and the final failure was witnessed. Therefore, accumulated DL of unidirectional laminates in spar cap and the following failure of shear web were considered to be the root cause of the blade failure.

5.2. The Brazier Effect

The Brazier effect was examined using the numerical model. The deformation of blade cross sections at the final failure are shown in Figure 17. The global flattening of these cross sections, which illustrates the Brazier effect, was not found in the deformed cross-sectional profiles possibly due to the constraint from shear webs between two spar caps, whereas the relative large out-of-plane deformation indicating local buckling was observed in the form of aft panel bulging outward and TE caving inward.

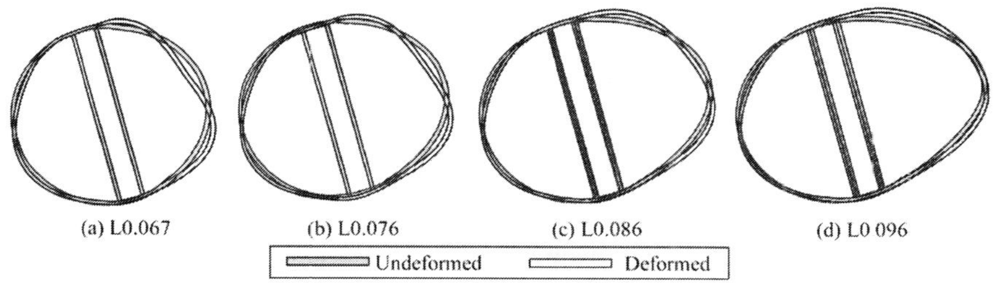

Figure 17. Deformation (scaled by 5) of cross-sectional profiles at the final failure. (**a**) L0.067; (**b**) L0.076; (**c**) L0.086; (**d**) L0.096.

The enlarged deformation of spar cap is shown in Figure 18, and it was found that the SS spar cap, which was initially convex in the transverse direction, became flattened and exhibited buckles where interfacial failure was significant. The relative out-of-plane deformation was defined as $(d_i - d_0)/d_0 \times 100\%$, where d_i and d_0 are the undeformed and deformed arc heights, respectively, of the spar cap at SS as shown in Figure 19.

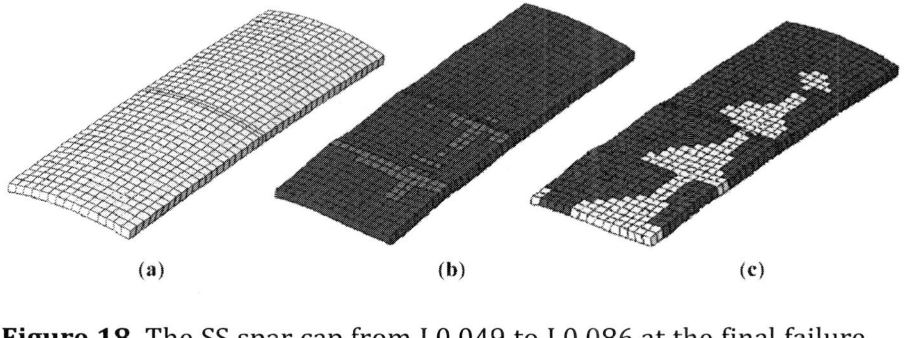

Figure 18. The SS spar cap from L0.049 to L0.086 at the final failure (deformation scaled by 10). (**a**) Undeformed; (**b**) Deformed with LF; (**c**) Deformed with DL.

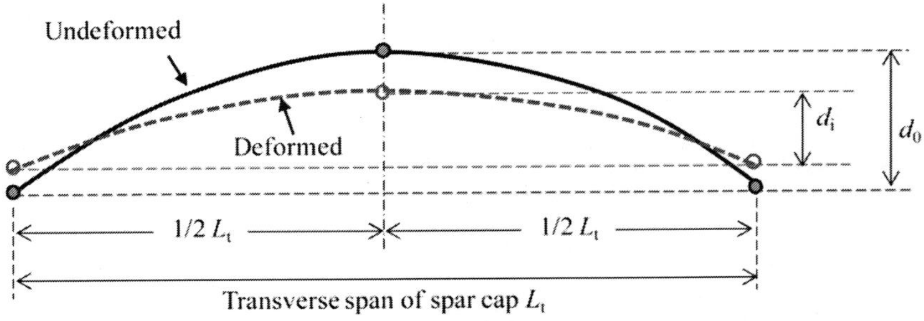

Figure 19. Parameters for the relative out-of-plane deformation of spar cap at SS.

Using this relative out-of-plane deformation, the local buckling response of spar cap was assessed quantitatively, and results at four locations of spar cap are shown in Figure 20, where initial slopes of the four curves were also plotted. It was observed that at certain load levels the curves of the relative out-of-plane deformation changed their trends and their nonlinearity became significant, and these load levels were determined as local buckling loads which had applied load fractions ranging approximately from 70% to 80%. The local buckling was expected to significantly increase out-of-plane deformation and through-thickness stresses which subsequently deteriorated DL of spar cap. Therefore, nonlinear out-of-plane deformation due to local buckling affected the failure process adversely and contributed to the blade failure.

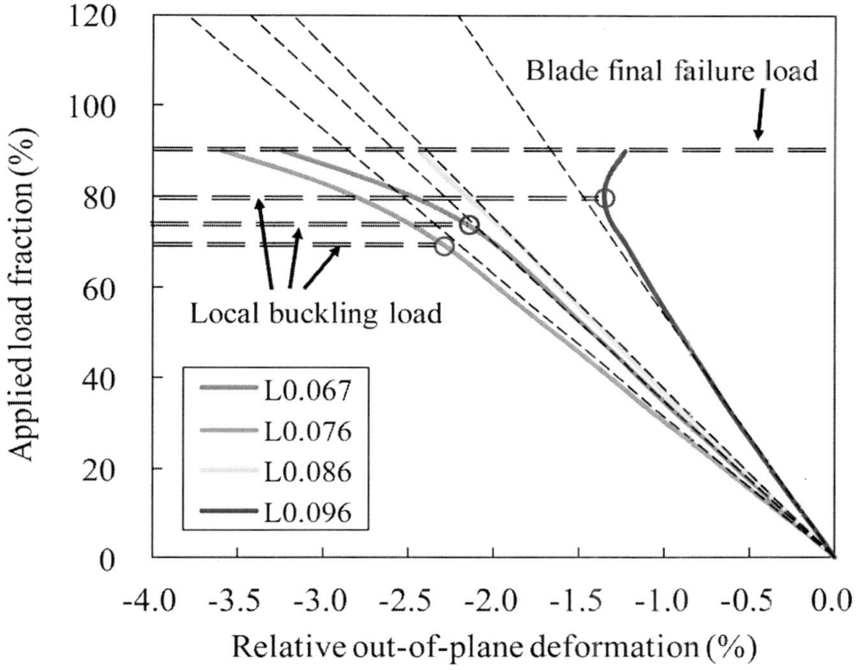

Figure 20. Relative out-of-plane deformation of the SS spar cap in the test case "F_Max_3.0".

6. CONCLUSIONS

This paper presented a thorough investigation into the failure of a 52.3 m composite wind turbine blade subjected to a series of static loads. Particular focus was placed on its complex failure characteristics observed in the test and a numerical model developed to predict failure behavior of the blade. Main conclusions can be drawn as follows: the blade failed catastrophically due to a total loss of load-carrying capacity under flap-wise bending. The final failure load was only 90% of the load once carried by the blade. Major failure was located at the transition region and characterized by multiple failure modes associated with both in-plane and more significant through-thickness stresses. Typical failure modes observed were delamination of the spar cap at the suction side, sandwich skin-core debonding, laminate fracture, and shear web failure. Among these failure modes spar cap delamination and shear web failure were indentified to be the root cause of the blade failure.

The loading history of the blade was numerically reproduced and the complex failure characteristics observed at the transition region were satisfactorily predicted in the FE simulation using the PFA techniques and a global-local modeling approach. It was found that the blade accumulated failure during its loading history. Local buckling contributed to the failure process by increasing the local out-of-plane deformation of spar cap, while the Brazier effect was not found in the transition region and it was not responsible for the failure of the blade.

This study emphasized the significance of through-thickness stresses on the failure of large composite blades. The proposed global-local modeling approach was proved to be effective to capture failure behavior of the blade while considerably reducing difficulties in modeling an entire blade.

ACKNOWLEDGMENTS

We are grateful to several anonymous reviewers for their constructive comments that have led to significant improvement of the paper.

CONFLICTS OF INTEREST

The authors declare no conflict of interest.

REFERENCES

1. Griffith, D.T. Technology Trends 2012. Presented at the 5th Sandia Wind Turbine Blade Workshop, Albuquerque, NM, USA, 30 May–1 June 2012.

2. Ashwill, T.; Laird, D. Concepts to Facilitate Very Large Blades. Proceedings of the 45th AIAA Aerospace Sciences Meeting and Exhibit, Reno, NV, USA, 8–11 January 2007.

3. Jensen, F.M.; Kling, A.; Sorensen, J.D. Change in Failure Type when Wind Turbien Baldes Scale-Up. Presented at the 5th Sandia Wind Turbine Blade Workshop, Albuquerque, NM, USA, 30 May–1 June 2012.

4. Jorgensen, E.R.; Borum, K.K.; McGugan, M.; Thomsen, C.L.; Jensen, F.M.; Debel, C.P.; Sorensen, B.T. Full Scale Testing of Wind Turbine Blade to Failure—Flapwise Loading; Riso-R-1392 (EN; Riso National Laboratory: Roskilde, Denmark, 2004.

5. Jensen, F.M.; Falzon, B.G.; Ankerson, J.; Stang, H. Structural testing and numerical simulation of a 34 m composite wind turbine blade. Compos. Struct. **2006**, 76, 52–61.

6. Jensen, F.M.; Weaver, P.M.; Cecchini, L.S.; Stang, H.; Nielsen, R.F. The Brazier effect in wind turbine blades and its influence on design. Wind Energy **2012**, 15, 319–333.

7. Jensen, F.M.; Puri, A.S.; Dear, J.P.; Branner, K.; Morris, A. Investigating the impact of non-linear geometrical effects on wind turbine blades—Part 1: Current status of design and test methods and future challenges in design optimization. Wind Energy **2011**, 14, 239–254.

8. Overgaard, L.C.T.; Lund, E.; Thomsen, O.T. Structural collapse of a wind turbine blade—Part A: Static test and equivalent single layered models. Compos. A **2010**, 41, 257–270.

9. Overgaard, L.C.T.; Lund, E. Structural collapse of a wind turbine blade—Part B: Progressive interlaminar failure models. Compos. A **2010**, 41, 271–283.

10. Yang, J.S.; Peng, C.Y.; Xiao, J.Y.; Zeng, J.C.; Xing, S.L.; Jin, J.T.; Deng, H. Structural investigation of composite wind turbine blade considering structural collapse in full-scale static tests. Compos. Struct. **2013**, 97, 15–29.

11. Chou, J.S.; Chiu, C.K.; Huang, I.K.; Chi, K.N. Failure analysis of wind turbine blade under critical wind loads. Engin. Fail. Anal. **2013**, 27, 99–118.

12. Facts, Analysis, Exposure to Industrial Wind Energy's Real Impacts. Avalilable online: http://www.windaction.org (accessed on 18 February 2014).

13. Marin, J.C.; Barroso, A.; Paris, F.; Canas, J. Study of damage and repair of blades of a 300 kW wind turbine. Energy **2008**, 33, 1068–1083.

14. Kong, C.; Bang, J.; Sugiyama, Y. Structural investigation of composite wind turbine blade considering various load cases and fatigue life. Energy **2005**, 30, 2101–2114.

15. Rajadurai, J.S.; Thanigaiyarasu, G. Structural analysis, failure prediction, and cost analysis of alternative material for composite turbine blades. Mech. Adv. Mater. Struct. **2009**, 16, 467–487.

16. Wind Turbines-Part 1: Design Requirements, 3rd ed.; IEC standard 61400-1; International Electrotechnical Commission (IEC): London, UK, 2005.

17. Guideline for the Certification of Wind Turbines; Germanischer Lloyd: Hamburg, Germany, 2010.

18. Sleight, D.W. Progressive Failure Analysis Methodology for Laminated Composite Structures; NASA/TP-1999-209107; Langley Research Center: Hampton, VA, USA, 1999.

19. Ridha, M.; Wang, C.H.; Chen, B.Y.; Tay, T.E. Modelling complex progressive failure in notched composite laminates with varying sizes and stacking sequences. Compos. A **2014**, 58, 16–23.

20. Anyfantis, K.N.; Tsouvalis, N.G. Post buckling progressive failure analysis of composite laminated stiffened panels. Appl. Compos. Mater. **2012**, 19, 219–236.

21. Samborsky, D.D.; Mandell, J.F.; Agastra, P. 3-D Static Elastic Constants and Strength Properties of a Glass/Epoxy Unidirectional Laminate. Available online: http://www.coe.montana.edu/composites/ (accessed on 27 November 2013).

22. Tsai, S.W.; Wu, E.M. A general theory of strength for anisotropic materials. J. Compos. Mater. **1971**, 5, 58–80.

23. Daniel, I.M.; Ishai, O. Engineering Mechanics of Composite Materials; Oxford University Press: Oxford, UK, 1994.

24. Christensen, R.M. The Theory of Material Failure; Oxford University Press: Oxford, UK, 2013.

25. Ye, L. Role of matrix resin in delamination onset and growth in composite laminates. Compos. Sci. Technol. **1988**, 33, 257–277.

26. Nagesh. Finite-element analysis of composite pressure vessels with progressive degradation. Def. Sci. J. **2003**, 53, 75–86.

27. Apalak, Z.G.; Apalak, M.K.; Genc, M.S. Progressive damage modeling of an adhesively bonded composite single lap joint under flexural loads at the mesoscale level. J. Reinf. Plast. Compos. **2007**, 26, 903–953.

28. Tsepres, K.I.; Papanikos, P.; Kermanidis, Th. A three dimensional progressive damage model for bolted joints in composite laminates subjected to tensile loading. Fatigue Fract. Engin. Mater. Struct. **2001**, 24, 663–676.

29. Dassault Systemes Simulia Corp. ABAQUS/Standard User's Manual, Version 6.12; Dassault Systèmes: Providence, RI, USA, 2012.

4

CHAPTER

FINITE-ELEMENT ANALYSIS OF PARTS STRESS STATE OF TIGHT JOINT ASSEMBLED BY PRESS FITTING

Sinitsyna Vasilya, Sinitsyn Anton

Mechatronic System Department, Izhevsk State Technical University, Izhevsk, Russia

ABSTRACT

This paper includes descriptions of the stress distribution regularities in the tight joint parts, regularities of the stress state changes in the contact region along coupling length, stress concentration factors, levels of additional stresses caused by press fitting. Distributions of stress intensity, axial stress, contact pressure, tangent stress in parts and in contact zone along coupling length are considered. Calculation results obtained by three approaches: Lame relationships, FEM without considering assembly method, FEM with considering press fitting process are analyzed and compared. The adequacy of research carried out is confirmed.

Keywords: Tight Joint, Stress State, Contact Pressure, Press Assembly, Finite Element Method

1. INTRODUCTION

The tight joints are widely used in the mechanical engineering. Increasing demands for higher transmitting forces and torques require the reliability enhancement of machinery and equipment. The tight joint reliability enhancement is one of the fundamental tasks as part of the problem reliability enhancement of machinery and equipment. The tight joint failure leads to machinery and equipment failure, which leads to technogeneous effect and economic troubles. Therefore, stress state of the tight joint parts should be calculated taking into account maximum number of acting loads to enhance the tight joints reliability.

Stress state of the tight joint parts depends on a lot of types of loads: loads, acting during component manufacture, assembly and operation. There are works [1] -[4] where the loads acting on components during mechanism operation are considered in stress state calculation. The tight joints can be assembled by heating the outer part, by shrinking the inner part, by press fitting, by hydraulic forging. Each of the joining method causes unique stress state of components.

In paper [5] parts stress state of the mechanically assembled tight joint is investigated. But, there is no consideration of distributions of stress intensity, axial stress, contact pressure, tangent stress in parts and in contact zone along coupling length. Stress concentration factors also are not considered.

The research is carried out with the financial support of RFBR in the framework of the research project *№* 14-08-31235*mol_a.* The press fitting is one the of mechanical assembly methods, whose classification is presented in the paper [6] . Scheme of press fitting is presented in Figure 1(b).

2. OBJECT OF RESEARCH

The shaft and hub connection with interference $N = 6$ мкм and coupling length $l = 10$ мм (Figure 1) is an object of research. Rated joint diameter $d = 10$ мм, inner diameter of shaft $d_1 = 4$ мм, outer diameter of hub $D = 30$ мм. Both parts are manufactured from the same material-steel 40, for which the yield point is $\sigma_T = 1480$ МПа, the strength is $\sigma_B = 1750$ МПа, the elastic modulus is $E = 2\times10^{11}$ МПа, the Poisson's ratio is $\mu = 0.3$, the density is $\rho = 7850$ кг/м^3. The friction ratio in the contact region is $f = 0.18$.

3. DESIGN METHODS

3.1. Analytic Approach

Lame dependences are often used for the initial evaluation of the tight joint parts stress state. According to them the largest radial tension is determined by the contact pressure.

$$\sigma_r = -p = -\frac{N}{\left(\frac{C_1}{E_1} + \frac{C_2}{E_2}\right)\cdot d} \tag{1}$$

where $C_1 = \frac{1+\left(\frac{d_1}{d}\right)^2}{1-\left(\frac{d_1}{d}\right)^2} - \mu_1$ and $C_2 = \frac{1+\left(\frac{d}{d_2}\right)^2}{1-\left(\frac{d}{d_2}\right)^2} + \mu_2$ -coefficient of stiffness; d-rated joint diameter; d_1-inner diameter of the shaft; d_2-outer diameter of the hub; N-rated interference, E_1, E_2, и μ_1, μ_2—elastic modulus and Poisson's ratio of the shaft and the hub respectively.

The largest compressive tangential stress in the shaft is evaluated by the expression:

$$\sigma_t^I = p\frac{2(d_1/d)^2}{1-(d_1/d)^2} \tag{2}$$

The largest compressive tangential stress in the shaft is estimated by the dependence:

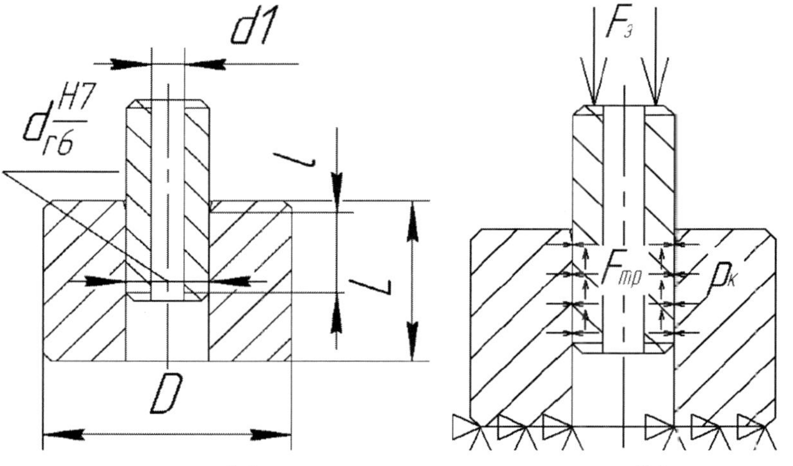

Figure 1. The tight joint (a) and the tight joints press fitting design model (b).

$$\sigma_t^{II} = p\frac{1+(d/d_2)^2}{1-(d/d_2)^2} \tag{3}$$

The largest equivalent stresses according Lame method are located on the inner surface of a hub:

$$\sigma_B^{II} = \sigma_t^{II} - \sigma_r \tag{4}$$

The considered Lame dependences are simple and convenient. However, they are received making assumptions that the mating parts have an ideal forms and excluding the phenomena and loads that are arising during the process of the tight joints assembly or operation.

3.2. Finite Element Method

Finite Element Method (FEM) allows solving the problem of determining the stress state of parts with the complex construction under various boundary conditions. To establish regularities of the stresses distribution in the contact region of mechanically as sembled couples it is necessary to solve two independent problems: the assembled tight joint modeling without taking into account assembly method, and modeling of the process of the shaft and the hub assembling by press fitting.

The components were modeled using solid elements. The contact was modeled using with Coulomb friction. For static modeling of the tight joint in the assembled state (Figure 2(a)) the grid consists from 453 axisymmetric element. For the tight joint modeling during the process of mechanical pressing the grid consists from 443 axisymmetric elements (Figure 2(b)). In supposed areas of stress concentrations and their gradients (contact surface and detail faces zones) the finite element mesh is a thickened (Figure 2).

Boundary conditions are represented by symmetry condition along Y axis and lower hub face fixing, that restrict all nodes displacement along X and Y axes. The 10 mm displacement of the shaft upper face is specified for the tight joint modeling during the process of mechanical pressing.

4. RESULTS

4.1. Results of Analytical Calculation

The results of analytical calculation of components stress state by means of Lame relationships are presented on Figure 3 in the form of radial and tangential stress distribution and in Table 1 in the form of numerical values.

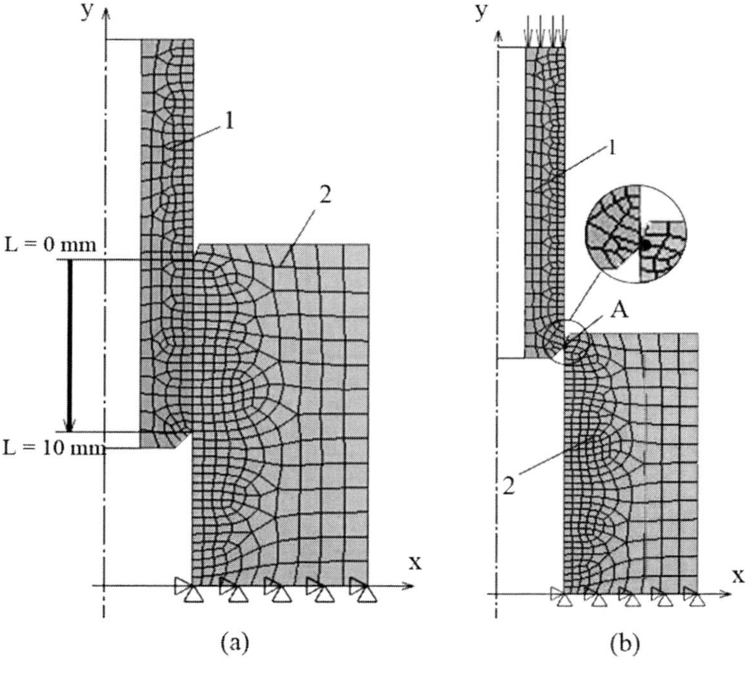

Figure 2. Finite element model of the tight joint: 1. the shaft, 2. the hub; A: characteristic point on the hub: (a) Static modeling of the tight joint in the assembled state; (b) The tight joint modeling during the process of mechanical pressing.

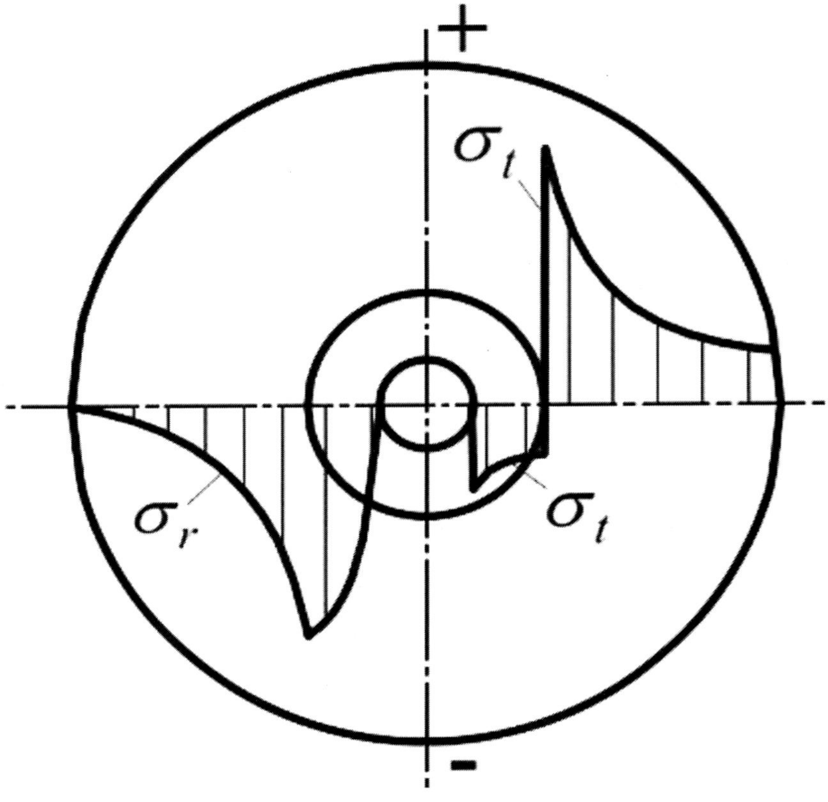

Figure 3. Radial and tangential stress distribution obtained by means of lame relationships.

Table 1. Results of stresses calculation by different method in the region of coupling length centre.

	Lame relationship	FEM without assembly method consideration	FEM with considering press fitting process
Radial stresses	45.61	45.86	47.38
Tangent stresses on the inner surface of the hub	57.01	51.73	51.11

The maximum radial stresses are located on the contact surface. The maximum tangential stresses of the shaft are shimming and located on the inner surface. The maximum shimming tangential stresses of the hub are tensile and located on the inner surface of the hub.

4.2. Results of FEM Calculation

4.2.1. Stress Intensity

Stress intensity distributions are commonly the same for calculation without considering assembly method and for calculation with considering press fitting process (Figures 4(a)-(b)).

However, in results of the calculation with considering press fitting process (Figure 4(b)) there is an area of low stresses (up to 14.98 MPa), that sharply comes to area of high (up to 82 MPa) stresses. The shaft stresses are lower than the hub stresses.

From Figure 4(b) one can see that, for the given initial condition the inner surface of the shaft is the more stressed (up to 128 MPa). Such stresses on the hub appear only near the coupling faces. Stress intensity in middle of the contact zone of hub inner surface varies around 100 MPa.

In the contact zone according to the length (Figure 5) the stress intensities of the covering part vary in the range from 93.45 MPa to 124.72 MPa for the calculation without considering assembly method and in the range from 95.07 MPa to 128.42 MPa for the calculation with considering press fitting process. Covered part in the contact zone has less stress state: the stress intensities vary according to the length in the range from 38.19 MPa to 91.41 MPa for the calculation without considering assembly method and in the range from 14.99 MPa to 95.93 MPa for the calculation with considering press fitting process.

The largest difference in numerical values of the stress intensity on the shaft and hub is observed at the distance of 0.50 mm from the top face of coupling. The phenomenon of wave formation on the chart of the stress intensity distribution on the shaft and hubwave (Figure 5, line 2) can be explained as a result of deformation that is arisedat the moment of the contact formation by the mating surfaces.

The additional stresses caused by mechanical assembly weaken the stress state of the shaft on the contact surface, and increase of the hub. The level of additional stresses caused by mechanical assembly reaches −62.38% of stress level calculated without considering assembly method on the shaft (Figure 5(a)) and 16.54% on the hub (Figure 5(b)).

For the calculation with considering press fitting stress intensity concentration factors in the faces zones get higher values then for the calculation without considering assembly method for both parts (Table 2). The maximum difference (44.09%) of concentration factors is observed near the top face of the hub.

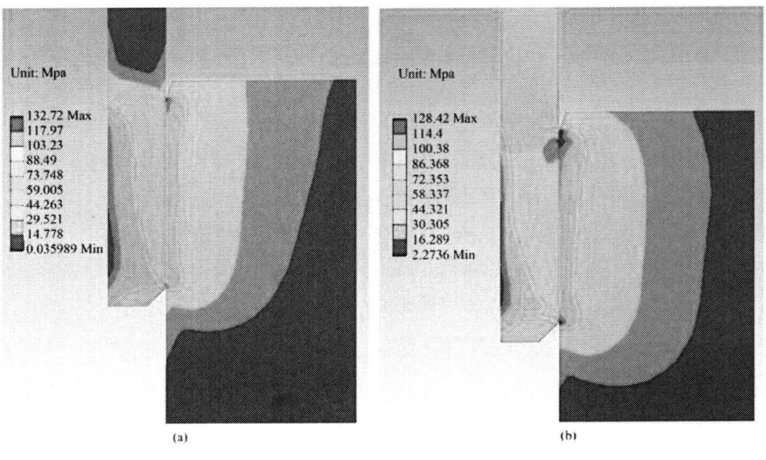

Figure 4. Stress intensity distributions: (a) Static modeling of the tight joint in the assembled state; (b) The tight joint modeling during the process of mechanical pressing.

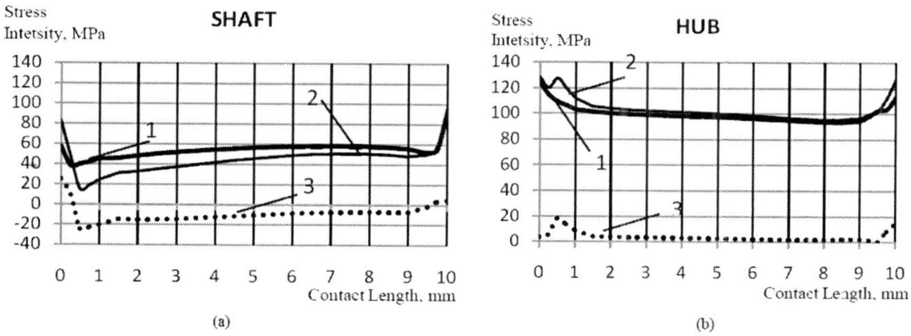

Figure 5. The stress intensity in contact zone along coupling length on the shaft (a) and hub (b): 1. The calculation without considering assembly method; 2. For the calculation with considering press fitting process; 3. The level of additional stresses caused by mechanical assembly.

Table 2. Stress intensity concentration factors.

		FEM without considering assembly method	FEM with considering press fitting process	Difference of Concentration factors
Shaft	Top face	1.27	1.83	44.09%
	Bottom face	2.04	2.14	0.49%
Hub	Top face	1.23	1.26	2.43%
	Bottom face	1.12	1.26	12.50%

4.2.2. Axial Stresses in the Contact Zone

Axial stress distribution in the tight joint parts provokes particular interest in consideration of longitudinal scheme of press fitting (Figure 6).

The axial stress distribution is more or less symmetry with respect to the contact length centre for static modeling of the tight joint in the assembled state. There is small nonsymmetry near the faces of the coupling, which is caused by different bevels. Thus, the maximum shimming stresses (48.9 MPa) are near the 45° bevel of bottom face of the coupling.

The transition of axial stress concentration zone from the bottom coupling face to top face is observed when comparing the results of two calculations (Figures 6(a)-(b). Maximum shrimping stresses (up to 116 MPa) are located on the shaft near the top face of coupling.

The greatest interest is the axial stress distribution directly in the contact zone, that is very different for calculation with and without assembly method consideration (Figure 7).

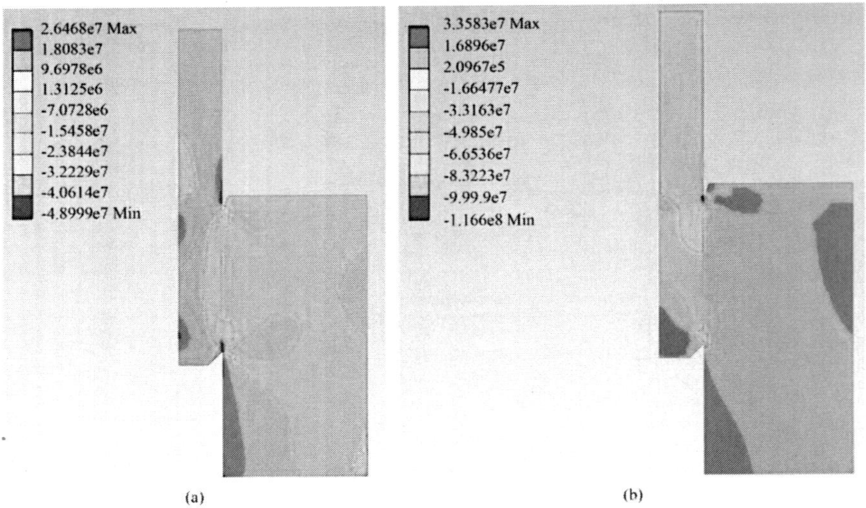

Figure 6. The axial stress distribution. (a) Static modeling of the tight joint in the assembled state; (b) The tight joint modeling during the process of mechanical pressing.

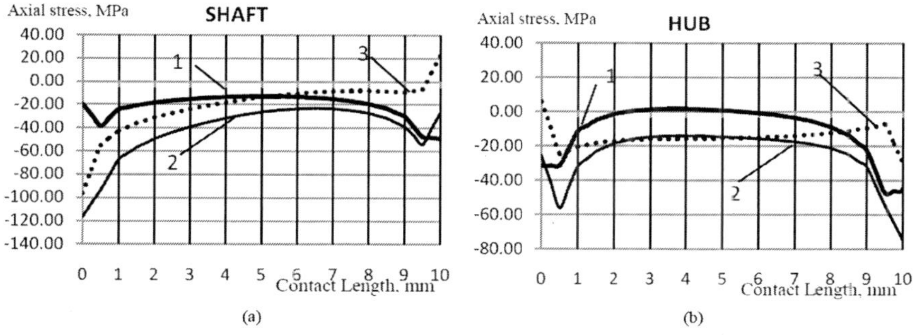

Figure 7. The axial stress in contact zone along coupling length on shaft (a) and hub (b): 1. The calculation without considering assembly method; 2. For the calculation with considering press fitting process; 3. The level of additional stresses caused by mechanical assembly.

On the hub maximum divergence of results (up to 29.6 MPa) is observed near the bottom face of the coupling and is 66% from values of axial stress calculated without assembly method consideration. On the shaft maximum level of additional stresses caused by mechanical assembly (up to 96.54 MPa) is observed near the top face of the coupling and is 481% from values of axial stress calculated without assembly method consideration.

Axial stress concentration factors near the faces of coupling has higher values when they are calculated with considering press fitting then when they are calculated without considering assembly method they are (Table 3). The maximum difference (205.43%) of concentration factors is observed near the top face of the hub.

4.2.3. Contact Pressure

Contact pressures have higher values when calculated with considering press fitting then when calculated without considering assembly method (Figure 8).

Maximum level of additional contact pressure caused by mechanical assembly (up to 40.58 MPa) is observed near the top face of the coupling and is 31.85% from values of contact pressure calculated without assembly method consideration.

Contact pressure concentration factors near the top face of the coupling are 2.88 and 2.42 for the calculation with considering press fitting process and the calculation without considering assembly method respectively (Table 4).

The maximum difference (19.08%) of concentration factors is observed near the top face of the hub.

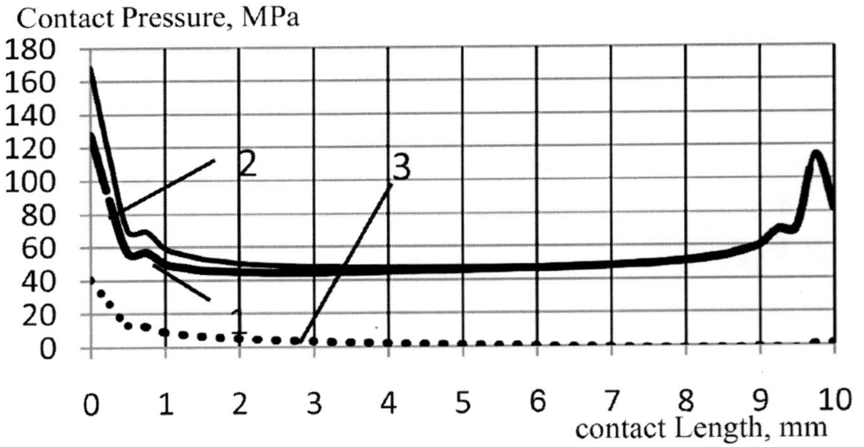

Figure 8. The contact pressure along coupling length: 1. The calculation without considering assembly method; 2. For the calculation with considering press fitting process; 3. The level of additional stresses caused by mechanical assembly.

Table 3. Axial stress concentration factors.

		FEM without considering assembly method	FEM with considering press fitting process	Difference of Concentration factors
Shaft	Top face	0.92	2.81	205.43%
	Bottom face	1.18	1.29	9.32%
Hub	Top face	1.27	2.26	77.95%
	Bottom face	1.90	3.01	58.42%

Table 4. Contact pressure concentration factors.

	FEM without considering assembly method	FEM with considering press fitting process	Difference of Concentration factors
Top face	2.42	2.88	19.08%
Bottom face	2.15	1.95	−9.30%

4.2.4. Tangent Stresses

Maximum divergence of results calculated with and without assembly method consideration is 41.18% on the hub and 54.67% on the shaft from values of tangent stresses calculated without assembly method consideration (Figure 9).

The additional tangent stresses caused by mechanical assembly (Figure 10, line 3) possess the negative values near the top face of the coupling and possess the positive values near the bottom face.

5. ADEQUACY OF CALCULATION RESULTS

The divergence of results calculated by Lame relationship and FEM without assembly method consideration in the region of coupling length centre is 0.5%. The divergence of the results in the region of coupling length centre calculated be Lame relationship and FEM with considering press fitting process is 3.8%. That confirms the adequacy of carried out research (Table 1).

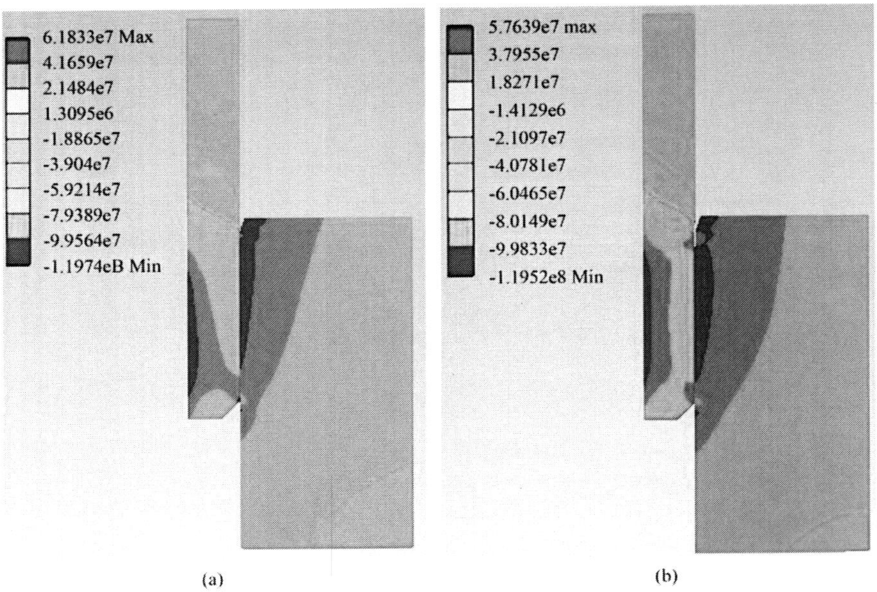

Figure 9. The tangent stress distribution. (a) Static modeling of the tight joint in the assembled state; (b) The tight joint modeling during the process of mechanical pressing.

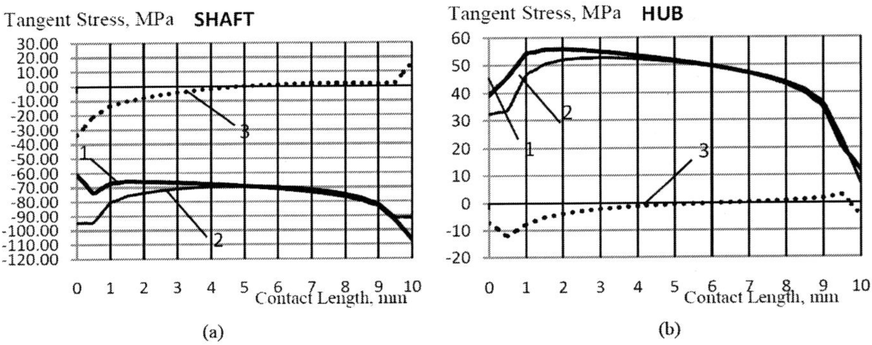

Figure 10. The tangent stresses on shaft (a) and hub (b): 1. The calculation without considering assembly method; 2. For the calculation with considering press fitting process; 3. The level of additional stresses caused by mechanical assembly.

6. CONCLUSIONS

1) Current research has found that mechanical assembly of the tight joint greatly affects the parts stress state. The level of additional stresses caused by mechanical assembly reaches −62.38% of stress level calculated without considering assembly method on the shaft. Stress concentration factors divergence is 44.09% near the top face of the shaft.
2) The tight joint parts stress state calculation with considering press-fitting process allows taking into account additional axial stresses arising in the course of mechanical assembly. Level of additional axial stresses in contact zone can reach 48.1% from values of axial stress calculated without assembly method consideration.
3) The tight joint parts stress state calculation with considering press fitting process allows estimating level of acting stresses and stress concentration factors more reliably because of taking into account additional stresses arising during assembly process. Therefore for tight joints reliability enhancement, the stress state calculation is recommended to do with considering assembly method.
4) The proposed approach to estimation of press-fitting influence on parts stress state is appropriate to use in their design and in CAD/CAM systems. That will lead to a more rational prescribing of geometrical dimensions of the mating parts and to the choice of their materials.

REFERENCES

1. Ast, M., Rosle, H. and Schenk, R. (1998) Finite Element Analysis of Shrink-Fit Shaft-Hub Connections. Voith Turbo Antriebstechnik GmbH, 1-15.

2. Abramov, I.V., Abramov, A.I., Sinitsyn, A.N. and Sinitsyna, V.V. (2010) Stress-Strain State of Tight Joints under a Bending Moment. Russian Engineering Research, 8, 759-762.

3. Abramov, I.V., Abramov, A.I., Sinitsyn, A.N., Sinitsyna, V.V. and Petrov, A.V. (2011) Experimental Plant for Inter- ference Fit Disassembling. 14th International Symposium on Mechatronika, 1-3 June 2011, Trencianske Teplice, 90- 92.

4. Abramov, I.V., Abramov, A.I., Sinitsyn, A.N. and Sinitsyna, V.V. (2011) Theoretical Study of New Methods of Assembly and Disassembly of the Tight Joints. Assembly in Mechanical and Instrument Engineering, 3, 11-15.

5. Abramov, I.V., Abramov, A.I., Sinitsyn, A.N. and Sinitsyna, V.V. (2014) Mechanical Assembly Method Influence on Tight Joint Parts Stress State. Mechanical Engineering Bulletin, 10.

6. Chenyatskiy, A.V., Sinitsyn, A.N. and Sinitsyna, V.V (2014) Approaches to Stress State Calculation of the Mechanically Assembled Tight Joint Parts. Bulletin of the Izhevsk State Technical University, 3.

5

CHAPTER

FINITE ELEMENT ANALYSIS OF IN-PLANE DISPLACEMENTS AND VON-MISES STRESSES IN ELLIPSOIDAL AND CIRCULAR CYLINDERICAL PETROLEUM TANKERS

Oluleke Oluwole*, **Eyere Emagbetere**

Mechanical Engineering Department, University of Ibadan, Ibadan, Nigeria

ABSTRACT

Road tankers are the most used means of transporting petroleum product to end users due to its cost effectiveness and energy-efficiency. The cylindrical tank has been well designed for by ASME VIII divisions 1 and 2 using analytical equations. Petrol tankers are not circular but elliptical probably for stability during transportation. This paper has used the finite element method to investigate in-plane displacements and Von-Mises stresses in both circular and elliptical cylindrical tanks under full loading. An elliptical OANDO® tanker of 66.78 m^3 volume and shell thickness of 0.2 mm and an equivalent volume circular cylindrical tank was used for the simulation. MATLAB® was used to generate geometrical mesh model of the petroleum tankers, extract element coordinates and conduct the finite element analysis. Plane strain condition was used in analyzing a section of the petroleum tanker. It was observed that an equivalent volume circular cylindrical tank was under a higher internal pressure (16,858 N/m^2) compared to the elliptical cylinder (14,480 N/m^2). Von-Mises stress and in-plane displacements showed direct linear relationships with internal fluid pressure. Von-Mises stress in the elliptical tank was found to be lower (5.7×10^6 N/m^2) than for the circular tank (8×10^6 N/m^2). In plane displacements was zero in the longitudinal direction for both tanks and of the order of 10^{-4} mm in the y-direction for both tanks with the circular larger by about 2.5×10^{-3} cm. So in addition to tank stability on the lorry, the Von-Mises stresses were lower as well for the elliptical tank. It was also observed that Von-Mises stresses were far below the yield stress of the steel plate. However, the effect of weldment area on lowering of yield stress was not studied. Stress values were validated using analytical method and found to be insignificantly different ($P > 0.05$).

Keywords: Von-Mises Stress; Petrol Tanks; Finite Element Analysis

1. INTRODUCTION

Road tankers are cylindrical or ellipsoidal pressure vessels used to convey liquids especially petroleum products. Petroleum tanker is widely used over other means of transporting petroleum products (pipeline, trains, badges and trucks) due to its cost effectiveness and energy-efficiency [1]. According to the projection of Ekpen [2], Nigeria's petroleum product consumption will rise to about 300 Kb/day, a large part ofher total energy consumption in the country. Recently, there has been increasing number of reports of tanker explosion during transit, claiming several lives and properties [3-9]. There is therefore need to consider petroleum tanker safety in line with design.

Road tankers constructed locallyare designed based on ASME standards for pressure vessels (ASME VIII divisions 1 and 2). The ASME standard presents design formulae that are simple to use, but limited to specific geometries and geometry details such as welded supports and openings. It does not put into consideration several actions or combination of actions such as local loads, seismic load, wind loads and external pressure in its design formula. A better approach is the design by analysis. According to Josef et al. [10], design by analysis is very flexible, it allows for any combination of actions, it can be used for complicated geometry, and it addresses directly the creativity of the designer. Finite element method is an efficient technique of design by analysis as it gives result close to corresponding experimental values.

The Finite Element Method is a numerical technique ideally suited to digital computers in which a continuous elastic structure (continuum) is divided (discretized) into smaller but finite well defined sub-structures (element) that can be represented by simple equations [11]. According to George [12], it represents an approximate numerical solution of a boundary value problem described by a differential equation.

There are several publications on application of Finite element analysis of specific pressure vessels. Mirko et al., [13] applied Finite element method in the determination of the stress state of a cylindrical vessel with variable thickness. The stress distribution of deformed tank i.e. a tank with part that has been formally repaired was analyzed [14]. Jaroslav [15], In his work made a listings of over 865 papers and conference proceedings on finite element in the analysis of pressure vessels and pipelines which he classified into the following categories: linear and nonlinear, static and dynamic;

stress and deflection analyses; stability problems; thermal problems; fracture mechanics problems; contact problems; fluid-structure interaction problems; manufacturing of pipes and tubes; welded pipes and pressure vessel components; development of special finite elements for pressure vessels and pipes; finite element software.

Most Finite element analysts develop own codes using MATLAB, C++, or FORTRAN. Meshing of geometrical models is done manually giving poor results. Other challenges faced by finite element analysts include: high cost of acquiring CAD software for numerical simulation, difficulty in managing read data from standard CAD programs such as dwg format (AutoCAD file) when using own code, and erroneous results, due to improper use of commercial software.

MATLAB is a computer language and the software has an interactive computing environment that enables numerical computation, analysis and data visualization [16]. This work has explored MATLAB's plotting capability and powerful computational tools to generate the geometrical mesh model of a petroleum tanker, extract element coordinates and conduct a finite element analysis of Von-Mises stress on a loaded tanker. The values obtained were compared with ASME standard.

2. METHODOLOGY

2.1. Automated Mesh Generator

The ellipsoidal tanker which is predominantly more common compared to the cylindrical ones has three inner segments (Figures 1 and 2). Each segment can be divided into four symmetrically similar sections (Figure 3). For ease and simplicity, the left bottom section was considered (Figure 4), and an automated mesh generation code was written with MATLAB®.

The common ellipse (Figure 5) and the ellipse Equation (1) was used in the development of the code. The automated mesh generator was developed using piecewise straight lines to approximate the curve and element continuity created during the meshing relations. The flowchart for the automated mesh generator is shown (Figure 6). Figure 7 shows meshing of the quarter section.

Figure 1. The tank model.

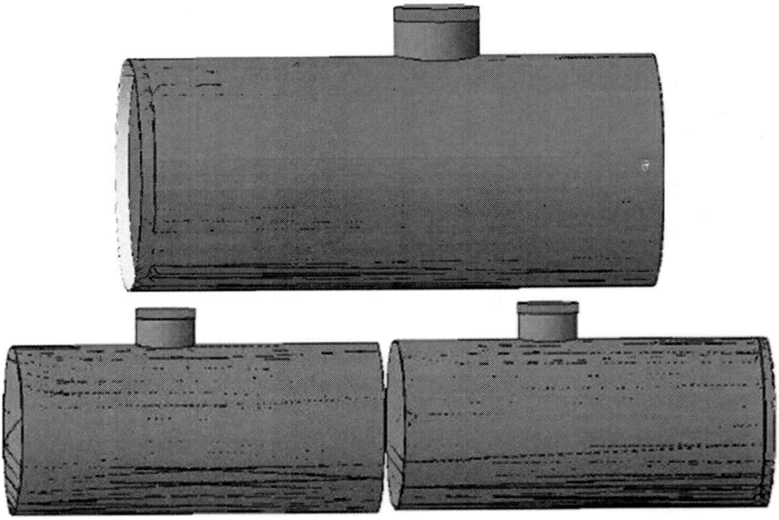

Figure 2. The tanker model divided into 3 parts initially joined together.

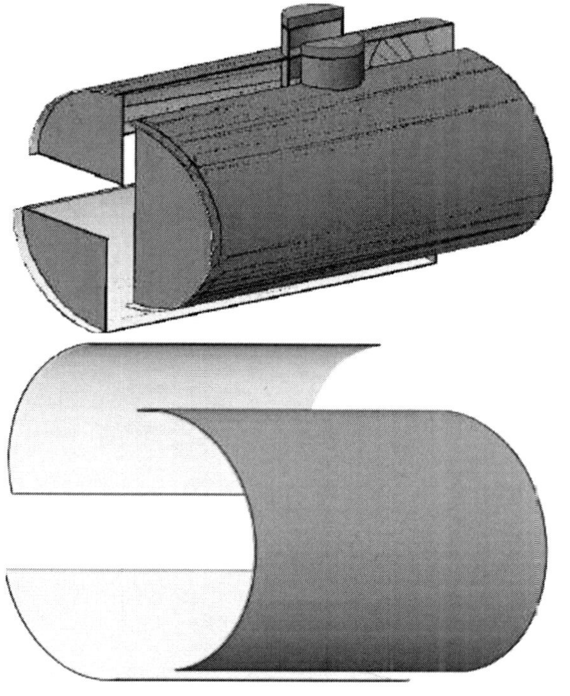

Figure 3. A part of the tanker further sectioned into parts.

Figure 4. Section for automated meshing.

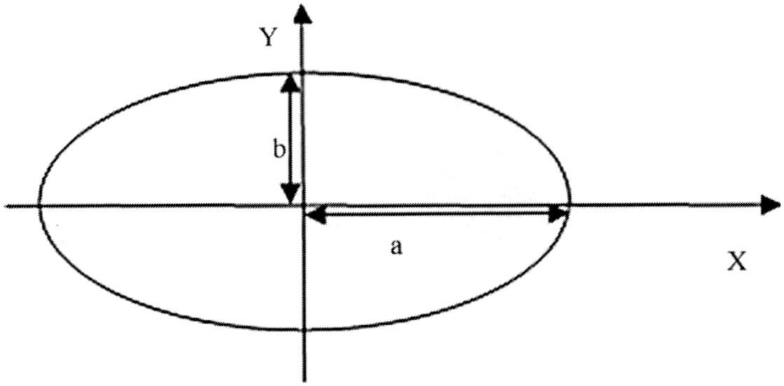

Figure 5. An ellipse with its dimension.

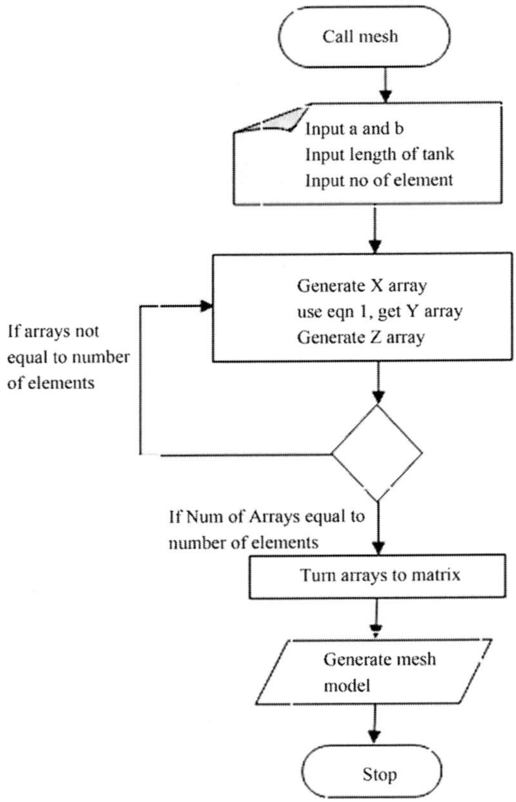

Figure 6. Algorithm for automatic mesh generator.

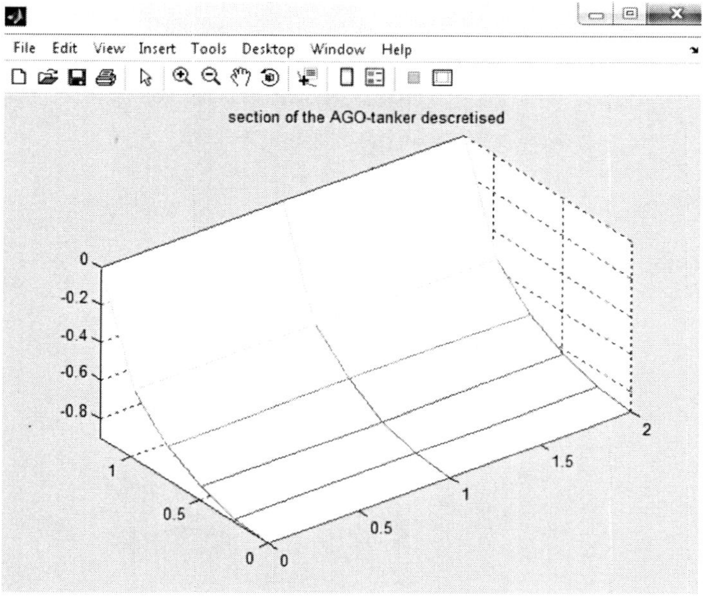
File Edit View Insert Tools Desktop Window Help
section of the AGO-tanker descretised
0
-0.2
-0.4
-0.6
-0.8
1
0.5
0
0
0.5
1
1.5
2

(a)

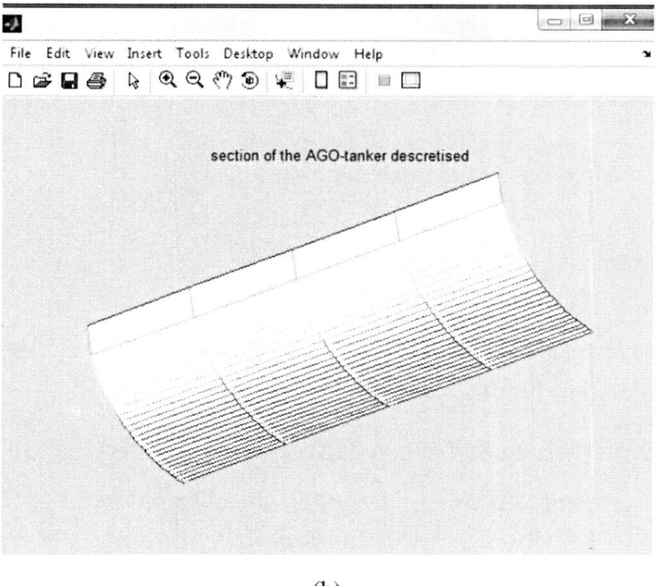
File Edit View Insert Tools Desktop Window Help
section of the AGO-tanker descretised

(b)

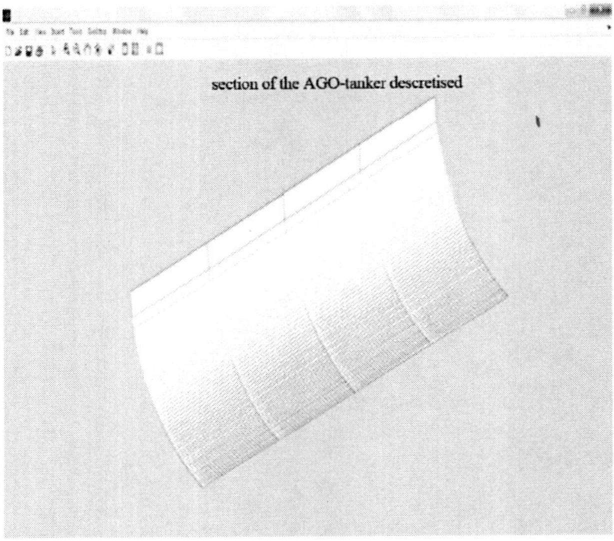

(c)

Figure 7. (a) 10 elements mesh; (b) 200 elements mesh; (c) 500 elements mesh.

$\frac{X^2}{a^2}+\frac{Y^2}{b^2}=1$ also was written as

$$Y=\sqrt{b^2-\frac{b^2\times X^2}{a^2}} \quad (1)$$

where a represents half the horizontal axis, and b half the vertical axis as shown in Figure 5.

2.2. Estimation of Stiffness Matrix and Force Vector

After generating the meshed model, an element was extrapolated and used in developing the stiffness matrix and force vector, needed to evaluate the displacement vector for plane strain elasticity of the model and VonMises stress.

The equation for the rectangular shell elements was derived by combining the linear rectangular element (membrane element) and the plate element (bending element).

2.2.1. Membrane Element

The equilibrium equations for a three-dimensional linear elasticity express in terms of stress is given [17,18] as;

$$\sigma_{ij,j} + f_i = 0$$

where $\sigma_{ij,j}$ are the stress components, f_i are the body forces.

The above equilibrium equation expressed in terms of displacement for two dimensional analyses is given as follows. The equation was reduced to two dimensional since plane strain condition is being considered.

$$-\frac{\partial}{\partial x}\left(A11\frac{\partial h_x}{dx} + A12\frac{\partial h_y}{\partial y}\right) - \frac{\partial}{\partial y}\left[A66\left(\frac{\partial h_x}{\partial y} + \frac{\partial h_y}{\partial x}\right)\right] = f_x$$

$$-\frac{\partial}{\partial x}\left(A21\frac{\partial h_x}{dx} + A22\frac{\partial h_y}{\partial y}\right) - \frac{\partial}{\partial y}\left[A66\left(\frac{\partial h_x}{\partial y} + \frac{\partial h_y}{\partial x}\right)\right] = f_y$$

h_x and h_y are the in-plane displacement in the x and y directions respectively of the Cartesian coordinate system.

2.2.2. Weak Formulation of the Membrane Element Equation

$$0 = \int_{y1}^{y2}\int_{x1}^{x2} t\left[\frac{\partial w_i}{\partial x}\left(A11\frac{\partial h_x}{\partial x} + A12\frac{\partial h_y}{\partial y}\right)\right.$$

$$\left. + A66\frac{\partial w_1}{\partial y}\left(\frac{\partial h_x}{\partial y} + \frac{\partial h_y}{\partial y}\right) + \rho w_{2h_x}\right]\mathrm{d}x\mathrm{d}y$$

$$-\int_{y1}^{x2}\int_{x1}^{x2} tw_{i,}f_x\mathrm{d}x\mathrm{d}y - \int_{re} tw_i t_x\mathrm{d}s$$

$$0 = \int_{y1}^{y2}\int_{x1}^{x2} t\left[\frac{\partial w_i}{\partial y}\left(A21\frac{\partial h_x}{\partial x} + A22\frac{\partial h_y}{\partial y}\right)\right.$$

$$\left. + A66\frac{\partial w_i}{\partial x}\left(\frac{\partial h_x}{\partial y} + \frac{\partial h_y}{\partial y}\right) + \rho w_{ih_y}\right]\mathrm{d}x\mathrm{d}y$$

$$-\int_{y1}^{y2}\int_{x1}^{x2} tw_i f_y \mathrm{d}x\mathrm{d}y - \int_{re} tw_i t_y \mathrm{d}s$$

2.2.3. Stiffness Matrix for the Membrane Element

The finite element equation is expressed as

$$[K]\{U\} = \{F\}$$

[K] is the stiffness matrix, {F} is the force vector and {U} is the displacement vector.

For plane strain condition, assuming homogeneity of material,

$$A11 = A22 = \frac{E(1-v)}{(1+v)(1-2v)};$$

$$A12 = A21 = vA11;\ A66 = \frac{E}{2(1+v)}$$

E and v are the material elastic constant and poison ratio respectively.

2.2.4. Interpolating Function for the Membrane Element

$h_i = [w_i]\{\mathbf{h}\}$ where $\{\mathbf{h}\}$ is the in-plane displacement vector, $[w_i]$ are the weight functions. The weight function for the rectangular element was used in the analysis.

2.2.5. Bending Element

The standard equation governing plate bending [19] is given as follows;

$$\frac{\partial^2}{\partial x^2}\left(B_{11}\frac{\partial^2\omega}{\partial x^2}+B_{12}\frac{\partial^2\omega}{\partial y^2}\right)+\frac{\partial^2}{\partial y^2}\left(B_{12}\frac{\partial^2\omega}{\partial x^2}+B_{22}\frac{\partial^2\omega}{\partial y^2}\right)$$

$$+2\frac{\partial^2}{\partial x\partial y}\left(2B_{66}\frac{\partial^2\omega}{\partial x\partial y}\right)=q$$

B_{ij} is the bending stiffness matrix for the shell element. E, v and t are the elastic modulus, poison ratio and thickness respectively of the tanker being considered.

$$B_{11}=\frac{Et^3}{12(1-v^2)};B_{22}=\frac{Et^3}{12(1-v^2)};$$

$$B_{12}=\frac{vEt^3}{12(1-v^2)};B_{66}=\frac{Gt^3}{12}$$

2.2.6. Weak Formulation of the Bending Element

$$0=\int_\Omega\left[\frac{\partial^2U}{\partial x^2}\left(B_{11}\frac{\partial^2\omega}{\partial x^2}+B_{12}\frac{\partial^2\omega}{\partial y^2}\right)\right.$$

$$+\frac{\partial^2U}{\partial y^2}\left(B_{12}\frac{\partial^2\omega}{\partial x^2}+B_{22}\frac{\partial^2\omega}{\partial y^2}\right)$$

$$\left.+4B_{66}\frac{\partial^2U}{\partial x\partial y}\frac{\partial^2\omega}{\partial x\partial y}-I_0v\right]\mathrm{d}x\mathrm{d}y$$

$$-\int_v\left[\left(\frac{\partial M_{xx}}{\partial x}+\frac{\partial M_{xy}}{\partial y}\right)n_x+\left(\frac{\partial M_{xy}}{\partial x}+\frac{\partial M_{yy}}{\partial y}\right)n_y\right]\mathrm{d}s$$

$$+\int_v\left[\frac{\partial U}{\partial x}\left(M_{xx}n_x+M_{xy}n_y\right)+\frac{\partial U}{\partial y}\left(M_{xy}n_x+M_{yy}n_y\right)\right]\mathrm{d}s$$

2.2.7. Stiffness Matrix for the Bending Element

For the plate element, on application of the classical plate theory [19], the Finite Element Model is given as follows

$$[K^e]\{\nabla^e\} = \{F^e\} + \{Q^e\}$$

Every term retains their meaning as above except for ∇ which is used to represent one out-of-plane displacement in the thickness axis and two rotations. Mass matrix was neglected since dynamics of the system was not considered.

2.2.8. Interpolation Function for the Bending Element

$\omega_i = [U_i]\{\omega\}$. Where $\{\omega\}$ represents one out-of-plane displacement in the thickness direction and two rotations and $[U_i]$ are the weight functions. B_{ij}' are the material elasticity matrices for the bending element.

$$B_{11} = \frac{E_1 t^3}{12(1-v^2)}, B_{22} = \frac{E_2 t^3}{12(1-v^2)}$$

$$B_{12} = \frac{v_{12} E_2 t^3}{12(1-v^2)}, B_{66} = \frac{G_{12t^3}}{12}$$

2.3. Calculation of p and q for Each Element

The stiffness matrices, and force vectors are basically a function of the side length of the rectangular element (Figures 8(a) and (b)) used to form the tanker model. From Figure 8, the values of p and q were calculated applying coordinate geometry as follows:

$$p = \sqrt{(y_4 - y_1)^2 + (x - x_1)^2}$$

$$q = z_2 - z_1$$

x_i, y_i and z_i are the coordinate locations of each element. I = 1 to 4 represents the four nodes of a 4 nodal rectangular element.

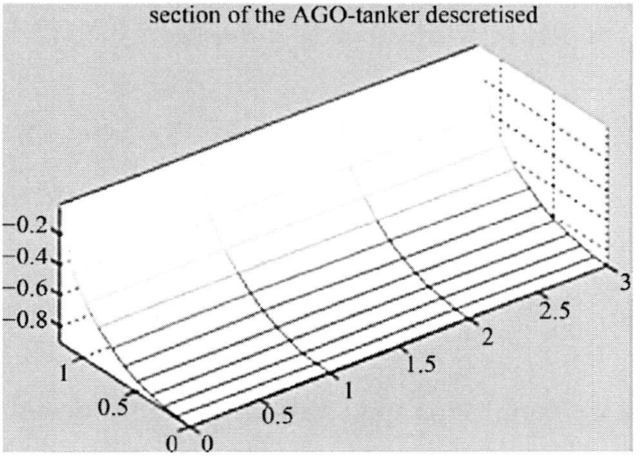

(a)

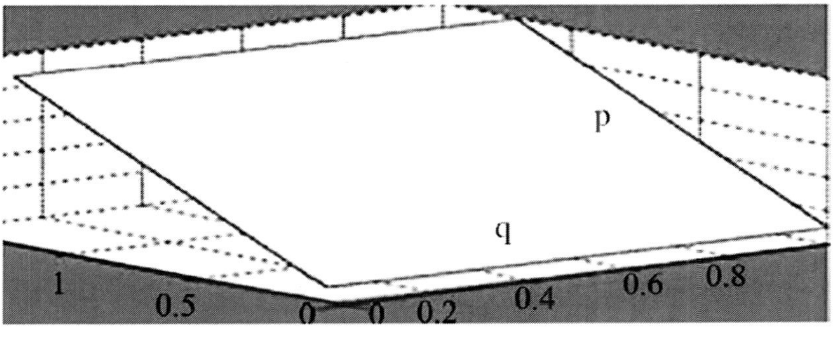

(b)

Figure 8. (a) Element mesh of the section; (b) An element showing dimensions for FEA.

2.4. Combination of Membrane and Bending Stiffness Matrices and Force Vectors

$$[K]+[Kb]=[KG]$$

where [K] is the 8 × 8 membrane stiffness matrix and [Kb] is the 12 × 12 bending stiffness matrix, both combining to give the total elemental stiffness matrix [KG] in the form shown below

$$[KG]=\begin{bmatrix} K & 0 \\ 0 & Kb \end{bmatrix}$$

The force vectors were combined in similar manner to form the total elemental force vector as follows

$$[FG]=\begin{Bmatrix} F \\ Fb \end{Bmatrix}$$

2.5. Boundary Conditions and Calculation of Displacement

Each element has five degrees of freedom (two in-plane displacements u and v, one out-of-plane displacements w, and two rotations. The displacements and bending at the edges were given zero since they are fixed.

Code for evaluating the finite element equation {U} $= [\mathbf{K}]/[\mathbf{f}]$, introducing boundary conditions and estimating the displacement was done using the flow algorithm (Figure 9).

2.6. Calculation of Stresses and Von-Mises Stress

$$[\text{Stress}]=[A]\times[B]\times\{U\}$$

A_{ij} represent the elasticity matrices for both the membrane element and bending element respectively.

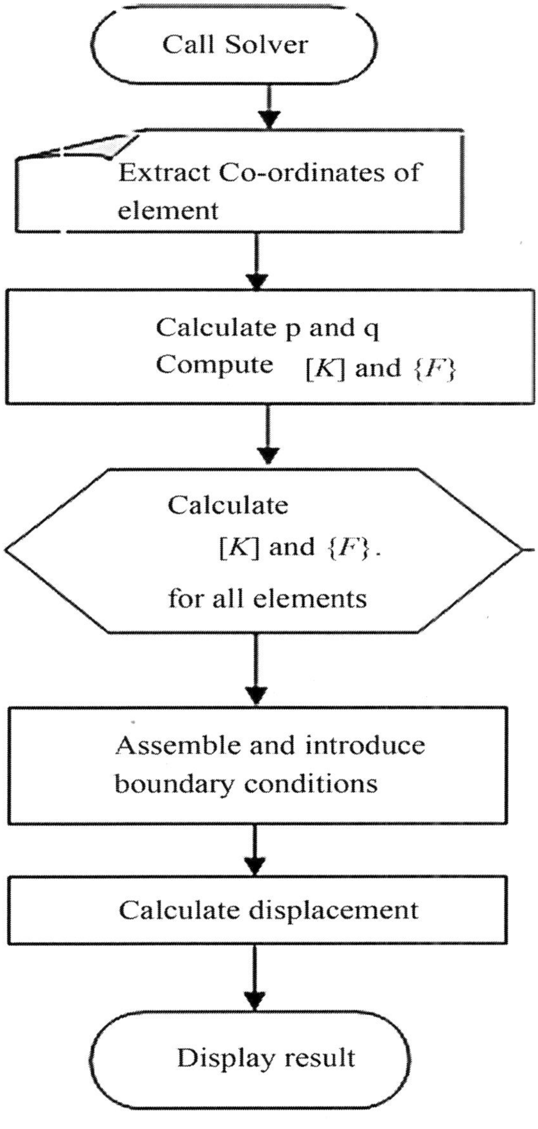

Figure 9. Algorithm applying extracted elemental coordinate for FE analysis.

[B] Represents the strain vectors and moment vector for membrane and bending element respectively.

[Stress] is the bending and membrane stress being computed.

2.7. Analytical Von-Mises Stress Calculation To compute the Von-Mises stress analytically, there are three principal stresses which are all functions of the pressure.

P1 = Principal stress 1 = Hoop stress = pressure × horizontal radii axis/t;

P2 = Principal stress 2 = Hoop stress/2;

P3 = Principal stress 3 = radial stress pressure;

$$\text{Von-Mises1} = \left(\sqrt{(P1-P2)^2+(P1-P3)^2+(P2-P3)^2}\right)/2$$

2.8. Dimensions Used for Analysis

Length of tanker = 485 cm;

Vertical axis of tanker = 180 cm;

Horizontal axis of tanker = 244 cm;

Thickness of tanker = 0.2 cm;

Poison ratio = 0.3;

Material of construction = A516M Grade 70;

Specified minimum yield stress = 25×10^7 N/m^2;

Maximum allowable stress = 13.8×10^7 N/m^2;

Elastic modulus = 200×10^9 N/m^2.

3. RESULTS AND DISCUSSIONS

3.1. Results

Figure 10 shows Von-Moses stress variation with loading. At full loading, Von-Mises stress was 5.7×10^6 N/m^2 and this decreased linearly to zero as the tank is offloaded until there is an empty tank with zero VonMises stress.

Figure 11 shows Von-Mises stress converge to a consistent value at large mesh elements for both elliptical and circular tanks. It also shows Von-Mises stress for circular tanks slightly higher than for the elliptical tank. Both FEM computed values converged to the analytical values for both circular and elliptical. ASME values tallied with the values for circular tank which are the same as the analytical values for the circular tank.

Figure 12 shows the effect of tanker internal pressure on in-plane displacement in the longitudinal (x) direction of elliptical cylindrical tankers. It shows a linear relationship between both parameters consistent with Hooke's law at elastic dimensions.

Figure 13 shows the effect of variation in internal pressure in elliptical tank on in-plane circumferential (y) displacements. It shows as well a linear relationship between both parameters.

Figure 14 shows the combined plots of in-plane longitudinal displacements for both circular and elliptical cylindrical tanks at full loading. The plots show that longitudinal displacements were the same for both tanks and were as it were non-existent (zero).

Figure 15 shows the combined plot of the circumferential in-plane displacements for both tanks at full loading. The plots show that the circumferential in-plane displacements were existent but at an order of 10^{-4} cm. The circular cylindrical tank showed higher circumferential displacement than the elliptical with a difference of about 2.5×10^{-3} cm.

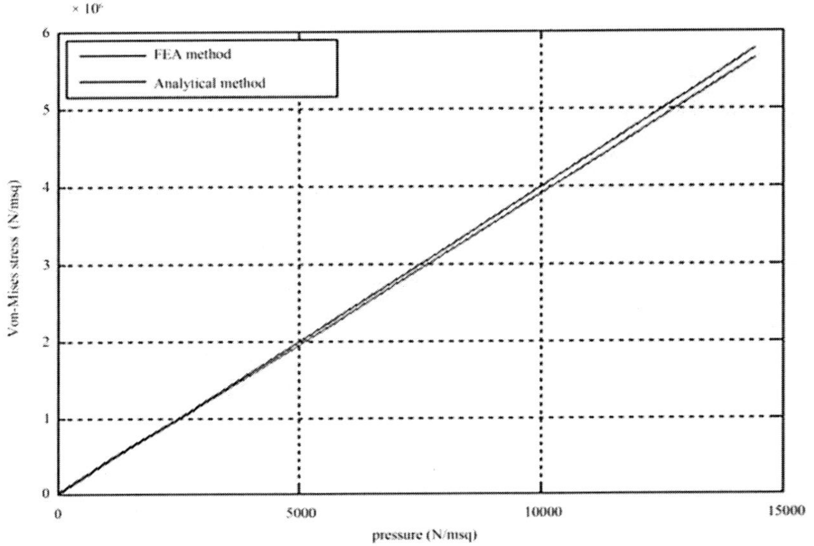

Figure 10. Von-Mises stress variation with internal tank pressure for elliptical cylindrical tank using 800 element meshes.

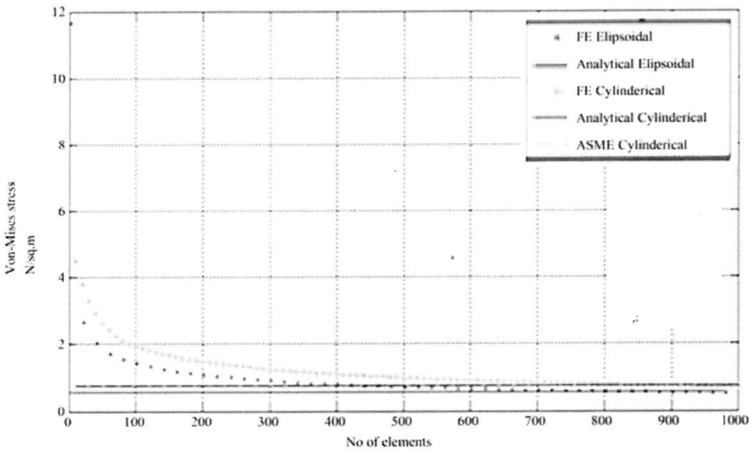

Figure 11. Von-Mises stress variation with no of meshing elements for cylindrical and elliptical cylindrical tank at 16,858 N/m^2 and 14,480 N/m^2 loading pressure respectively.

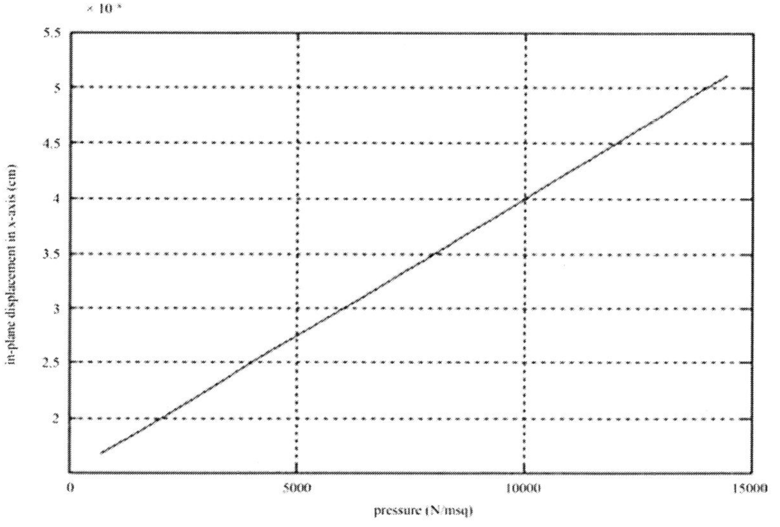

Figure 12. Variation of in-plane displacement in the longitudinal (x) direction of elliptical cylindrical tankers with internal load pressure.

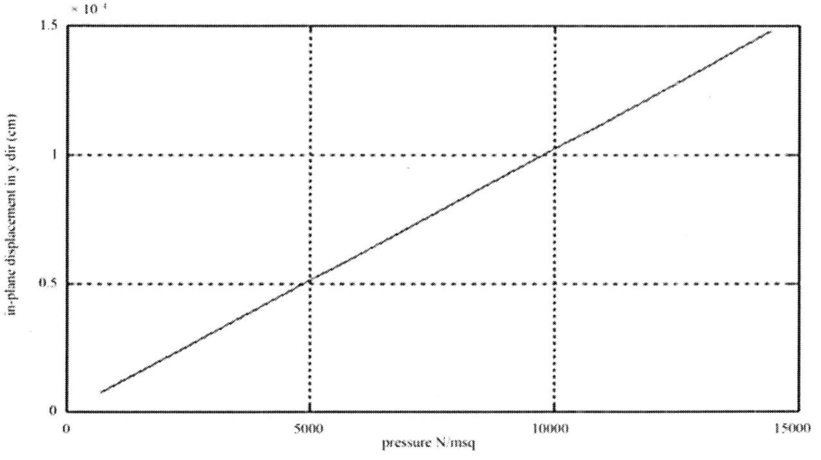

Figure 13. Variation of in-plane displacement in the circumferential (y) direction of Elliptical cylindrical tankers with internal load pressure.

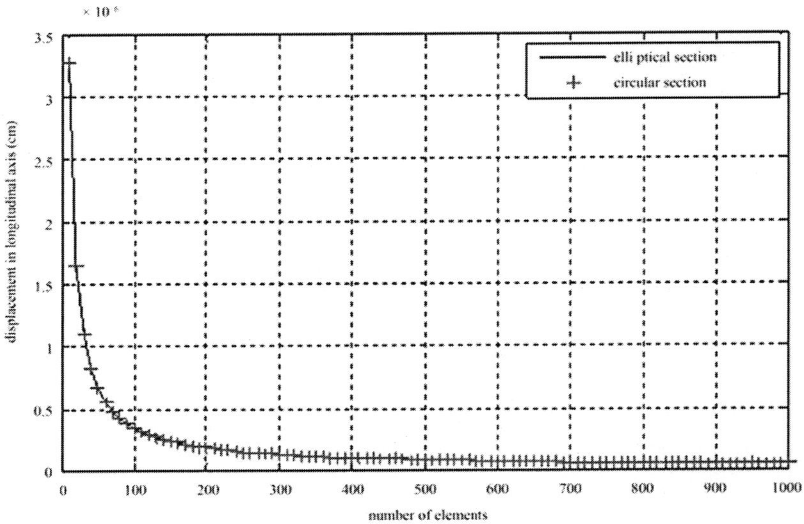

Figure 14. In plane displacement in the longitudinal (x) directionof elliptical and circular cylindrical tankers at full loading.

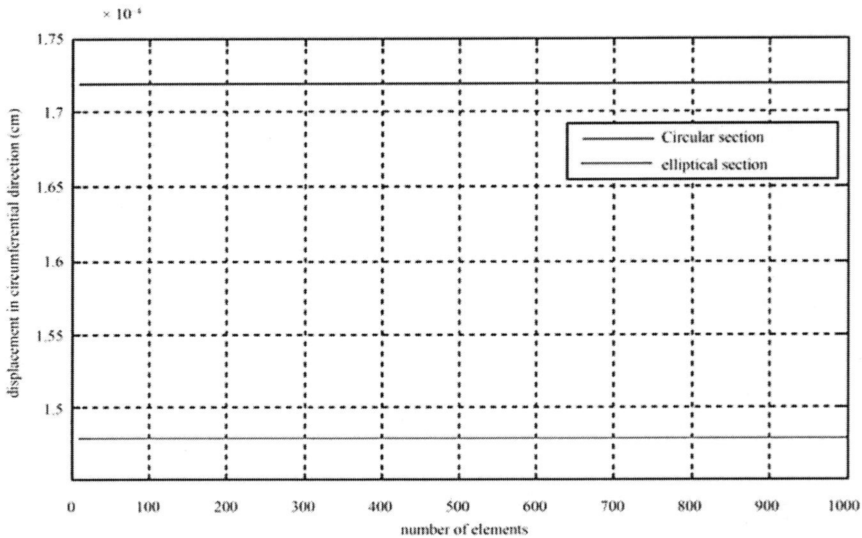

Figure 15. In plane displacement in the circumferential (y) direction of elliptical and circular cylindrical tankers at full loading.

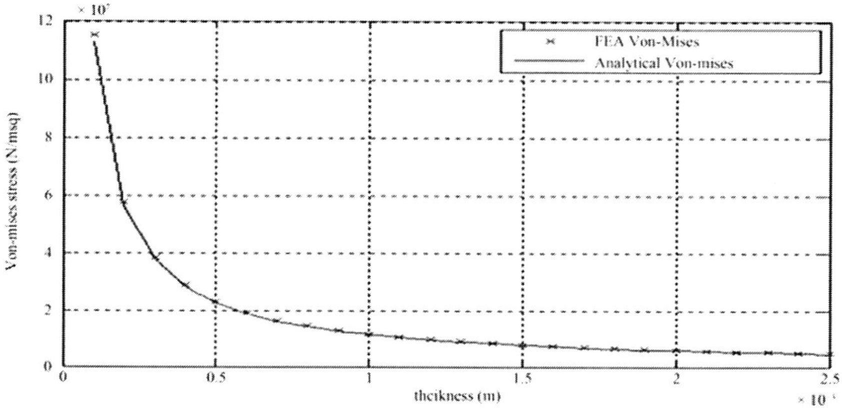

Figure 16. Von-Mises variation with tanker shell thickness.

Figure 16 shows Von-Mises stress variation with tanker shell thickness. It shows an increasing stress as shell thickness is reduced and after 5.0×10^{-2} m shell thickness, Von-Mises stress increased rapidly till yield stress is reached.

3.2. Discussion of Results

The direct linear relationships between internal loading pressure and Von-Mises stress (Figure 10) and that between in-plane displacements and internal pressure (Figures 12 and 13) are not unexpected because of the linearity of the stress-strain relationship in the elastic region. The result clearly showed that at full load Von-Mises stress is maximum (Figures 10 and 11). However, the Von-Mises stress for the fully loaded tanker still fell far below the yield stress of the material and the maximum allowable stress (Figure 11). This showed that the tanker will not yield during transportation from the loading pressure. However, since the tanks are always welded and not seamless, the tanks are susceptible to yield stress lowering effect of heat-affected-zone (HAZ) from welding. The HAZ causes embrittlement, hence a lowering of yield stress and if care is not taken, there could be microhole defects from where rapid crack propagation could develop causing catastrophic failure of the tank. If this effect is coupled to the corrosive effect of the petroleum product over time causing what can be termed stresscorrosion effect on the welded portion could as well cause failure over time. However, these issues were not addressed in this paper and are concerns for future work.

The results also show that circular cylinderical tanks have higher internal pressures as well as higher VonMises stresses than elliptical tanks of the same loading volume (Figure 13). The lower Von-Mises stresses developed in elliptical cylindrical tanks as well as their stability during transportation gives the elliptical tank a big edge over the circular cylindrical tank.

The in-plane displacements (Figures 14 and 15) showed good tanker material integrity and nullifies any idea of failure by yielding but just as had been mentioned earlier future work on weldment defects coupled with stresscorrosion need to be done to investigate what is transpiring here which could lead to other types of failure over time.

The inverse and non-linear relationship between tank shell thickness and Von-Mises stress (Figure 16) was also not surprising because of the relations between hoop/longitudinal stresses and shell thickness. This result showed clearly that it is important to design for a thickness that would guarantee a sufficient factor of safety especially since the effect of the welded parts' heat affected zone (HAZ) on lowering of yield stress must be considered.

4. CONCLUSION

Von-Mises stresses in circular and elliptical petroleum tankers under full loading have been obtained using the finite element method. It was observed that an equivalent volume circular cylindrical has a higher internal pressure (16,858 N/m^2) compared to the elliptical cylinder (14,480 N/m^2). Von-Mises stresses showed a linear relationship with variation in internal fluid pressure. Von-Mises stresses in the elliptical tank were found to be lower (5.7 × 10^6 N/m^2) than for the circular tank (8 × 10^6 N/m^2. VonMises stresses were far below the yield stress of the steel plate (25 × 10^7 N/m^2). However, the effect of weldment area and stress corrosion on lowering of yield stress was not studied. Values obtained for the circular cylindrical tank were compared with ASME VIII divisions 1 and 2 standard values and found not to be significantly different ($P > 0.05$).

REFERENCES

1. KSP 10-K, "Transportation of Petroleum Products," 2012. www.wikinvest.com/stock/KSea_ Transportation_Partners_ LP_%28KSP%29/Transportation

2. J. O. Ekpen, "Oil Refinery in Nigeria. (Myths and Truths)," NAEE Conference, Abuja, 28-29 April 2011, 2012. http:// naee.org.ng/files/

3. K. Babasoki, "Abuja Tanker Explosion," 2012. Fromogala. wrodpress.com/2011/11/16/latest-news-fromthe-abuja-tanker explosion

4. A. Ola and O. David, "10 Killed in Multiple Tanker Explosions, Building Collapse," 2012. wwww.vanguardngr. com/ 2012/06/10-killed-in multiple-tanker-explosion-building-collapse

5. Daily Mail Reporter, "86,000 Gallon Oil Tanker Bursts into Flames in Highway Crash Caused by Drunken Driver," 2012. www.wdailymail.co.uk/news/article-2126891/fuel-tanker-bursts-into-flames-us-freeway-crash-caused-by-drunken-driver

6. M. Emeka, "Ogun State Government Silent on Fuel Tanker Explosion," Vanguard, 2012, 6 Pages.

7. O. Uchechukwu and A. Boila, "Vehicles Burnt at Ibadan Tanker Explosion," 2012. www.tribune.com/ng/.../37648-12vehiclesburnt-inibadan-tanker

8. C. Miley, "Melborne Tanker Explosion Burns 9," 2012. news ninemsn.com.all/ national/8387836/petrol-tanker-explodes -in -melborne-suburb

9. B. Godfey and O. Ifeyinwa, "Oil Tanker Explosion Destroys Containers, Cars at Apapa Port," 2012. www.vanguardngr .com/2012/04/oil-tanker explosion-destroys-containers-cars - at-apapa-port

10. L. Z. Josef, R. Franz and S. Sebastian, "Pressure Vessel Design—The Direct Route," 1st Edition' Elsevier ltd., Oxford, 2006.

11. G. B. Richard, "Advance Strength and Applied Stress Analysis," 2nd Edition, McGraw Hill Publishing Companies Inc., New York, 1999.

12. R. B. George, 1995. "Finite Element Analysis," Shaum's Outline Series, McGraw-Hill, New York, 1995, 73 Pages.

13. Đ. Mirko, P. Dragan and B. Milan, "Identification of the Stress-Strain State of a Cylindrical Tank with Walls of Variable Thickness," FME Transactions, Vol. 39, No. 1, 2011, pp. 25-32.

14. V. P. Anatoly, Y. F. Sergiy, S. K. Viktor and Z. K. Edward, "Analysis of the Stressed State of Cylindrical Tanks with Defects of Geometrical Shape," SCAD Group, Kiev, 2003.

15. M. Jaroslav, 1996. "Finite Elements in the Analysis of Pressure Vessels and Piping—A Bibliography (1976- 1996),"

International Journal of Pressure Vessels and Pipings, Vol. 69, No. 3, 1996, pp. 273-339.

16. M. Patrick and A. O. Thomos, "Graphics and GUIs with Matlab," Chapman & Hall CRC Press Company, London, 2003.

17. T. Timoshenko and J. Goodier, "Theory of Elasticity," McGraw Hill Books Company Inc., New York, 1951.

18. C. Ross, "Advance Stress Analysis," Ellis Horwood Limited, New York, 1987.

19. J. N. Ready, "An Introduction to Finite Element Method," 3rd Edition, McGraw Hill, New York, 2006.

6

CHAPTER

FINITE ELEMENT ANALYSIS COUPLED WITH FEEDBACK CONTROL FOR DYNAMICS OF METAL PUSHING V-BELT CVT

Toshihiro Saito

[1] Honda R&D Co., Ltd. Automobile R&D Center, , Japan

1. INTRODUCTION

Nowadays automobiles are required to meet environmental requirements, such as lower exhaust emissions and higher fuel economy. One of the key factors for improving the overall efficiency of a vehicle is the efficiency of its transmission.

A CVT has a greater potential for improving fuel economy than a step-type automatic transmission (AT), because of its integrated control with the engine [1]. That is, CVTs are capable of continuously tracing engine operating ranges with high fuel efficiency. Another advantage is that CVTs allow vehicles to drive without lowering the driving torque or the engine rpm while shifting the gear ratio.

However, when the transmission efficiency of a CVT by itself is compared with a step-type AT, CVT is known to have lower efficiency because its driving torque is transferred by means of contact and friction [2]. The transmission efficiency of a CVT is determined by friction loss at its oil pump and metal pushing V-belt. The oil pump must produce enough pulley pressure so that the metal V-belt mounted between two pulleys does not slip. A higher pulley pressure, however, means a greater friction loss at the oil pump [3].

As for the metal V-belt, gradually lowering pulley pressure while maintaining a constant transmission torque increases the transmission efficiency of the belt by itself, as long as it does not slip on the pulleys. However, the transmission efficiency begins to drop under a certain operating condition. This implies the existence of an optimum operating condition that maximizes the transmission efficiency of the belt [4]. To find this condition, it is important to predict friction loss at each portion of the metal V-belt during CVT operation.

A considerable amount of research has so far been made on methods for calculating friction loss that occurs at each part of the V-belt, but many of them use simple equations that are based on assumptions and not linked with dynamic belt behavior [5]. Accordingly, although these methods simulate the tendency of friction loss on each belt part, they are not sufficient for examining the influence on friction loss of metal belt shape and pulley rigidity.

A commercially available multi-body simulation (MBS) code used to be chosen for the dynamic belt behavior [7]. In the MBS model, each V-belt block was treated as a separate rigid body. Contact normal forces were modeled between adjacent blocks, using realistic stiffness and damping properties. Contact normal forces, along with both radial and circumferential friction forces, were also modeled between the block edges and pulley faces. Block geometry and material properties were used to arrive at contact stiffness and damping characteristics. As such, the MBS model contained a relatively small number of degrees of freedom. At the results, the contact forces between the block edges and pulley faces could be only obtained, not the contact pressure distribution.

Therefore, the following describes how CVT-ratio control logic was included in the previously reported technology for simulating the dynamic behavior of a metal pushing V-belt, and presents how the new simulation can closely reproduce actual operation. Thus, this paper reports a technology developed to predict the transmission efficiency of a CVT drive system comprising a metal pushing V-belt and pulleys.

2. DEVELOPMENT AIMS

2.1. Maintaining speed ratio with feedback control

Figure 1 shows the main section of the CVT used in this development. A metal V-belt is mounted in the V groove on two pulley shafts, and a pair of movable pulleys are mounted on the shafts to face each other. The movable pulleys are shifted in the axial direction by line oil pressure supplied from the inside of the shaft. When the CVT is in operation, feedback control is exercised to maintain an arbitrary speed ratio between the two pulleys, varying the oil pressure applied to the large-diameter (in terms of belt mounting position) pulley while maintaining a constant pressure for the small-diameter pulley. When taking an example of the top ratio, the drive pulley speed, driven pulley torque, driven pulley oil pressure, and target ratio are input to the feedback control system. The system then outputs the driven pulley speed, drive pulley torque, and drive pulley oil pressure. The input-output relation differs from the metal V-belt behavior simulation technology developed previously, so it was necessary to incorporate a new feedback control into the simulation in this project.

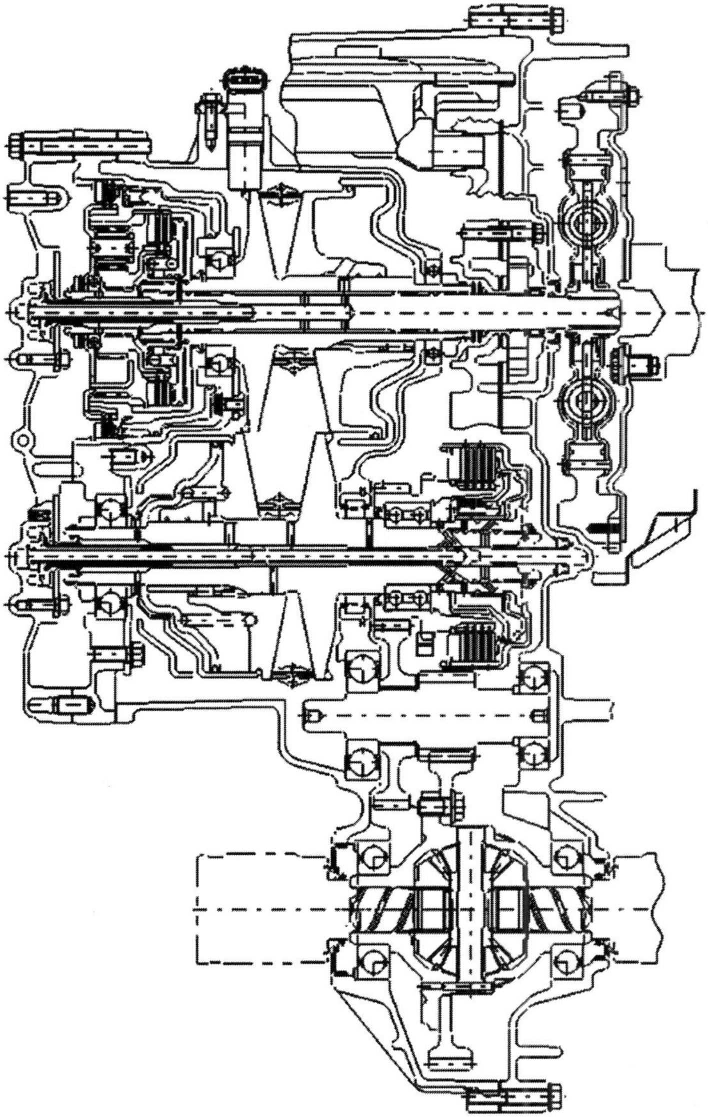

Figure 1. CVT cross section

2.2. Pulley shaft thrust load ratio

Pulley shaft thrust load is obtained as the sum of two values: the product of oil pump line pressure and its acting area, as well as the product of centrifugal oil pressure and its acting area. The ratio between drive pulley thrust load and driven pulley thrust load, referred to as the pulley shaft thrust load ratio, has a positive correlation with the speed ratio; therefore, this ratio is uniquely determined once the speed ratio is set. Because the developed simulation outputs pulley shaft thrust loads, its prediction accuracy can be verified by comparing simulated and measured values.

2.3. Transmission efficiency

Transmission efficiency can be obtained by multiplying the ratio between drive shaft torque and driven shaft torque by the speed ratio, as expressed in Equation (1). Here, each torque value can be obtained as the product of a tangential friction force, generated between the V-face of each pulley and metal V-belt elements, and the effective element V-surface radius. As such, it is necessary to accurately predict the direction of a friction force acting on each element V-surface and its effective radius under that condition. The effective radius of the element V-surface is influenced by its contact pressure distribution. This means that it is also necessary to consider the elastic deformation of the element V-surface.

For this reason, to simulate the transmission efficiency of the metal pushing V-belt, the following were set as development objectives:

1. Implement feedback control equivalent to speed ratio control performed in an actual vehicle

2. Quantify sliding velocities and friction forced at various element contact areas

3. Calculate friction losses at various portions of the metal V-belt

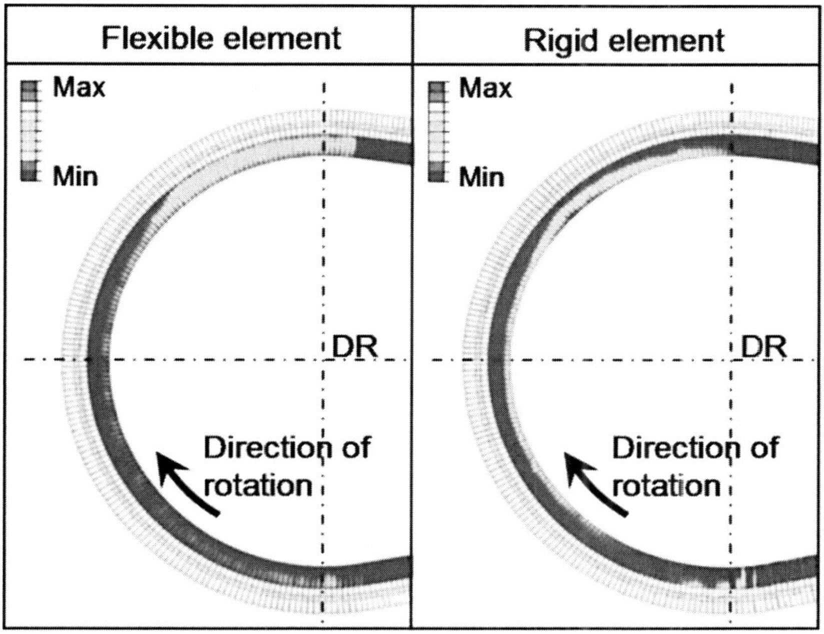

Figure 2. Comparison of friction force distribution of Element V-surfaces on DR pulley

3. TECHNOLOGY OF PREDICT TRANSMISSION EFFICIENCY OF METAL PUSHING V-BELT

3.1. V-belt model consideration of element deformation

The transmission efficiency of a metal pushing V-belt is mainly determined by friction losses that occur between its elements and pulleys and between the elements and rings. Friction loss has a correlation with the product of the friction force and sliding velocity of a friction surface. Accordingly, a key to predicting the transmission efficiency is to simulate the friction forces and sliding velocities of the element V surfaces in contact with the pulley. These friction forces and sliding velocities have distributions on the pulley, and these distributions cannot be simulated accurately unless element deformation is taken into account (Figure 2). Because the effective radius of elements would be influenced by the contact pressure distribution. For this reason, an element stress prediction model [6] previously designed to consider element deformation was modified to make a model of the metal V-belt used in this study.

Figure 3 and Figure 4 show the model. Two flexible solid elements are employed in the direction of thickness for individual elements to reproduce bending deformation. Three flexible solid elements are used to divide the R section of the element neck, where stress is thought to concentrate. Reproduction of detailed form of the nose and hole sections has been prioritized, and therefore modeled with rigid elements. Four flexible shell elements have been used in the direction of width to model each layer of rings to enable reproduction of the crowning of the rings. In addition, the ring-element, element-element, ring-ring and element-pulley contact surfaces have been defined, and appropriate friction characteristics have been assigned in each case.

To make the load conditions around the belt pulleys closely resemble the layout in an actual CVT, beam elements were used to express the shafts. In this model, the shafts are supported at the bearing positions. Also, a gear is provided to mesh with the driven (DN) shaft to reflect reaction forces applied by the gear. Because the belt mounting diameter varies depending on speed ratios, deflection rigidity calculated for the mounting positions of each ratio was applied to the pulley V-face. Regarding the relation between each shaft and the movable pulley, the model defines a fitting clearance at their engagement position, as well as a backlash in the rotational direction at the roller position.

Figure 5 shows the flow of analyzing the metal V-belt. In this flow, the belt is initially placed at the perfect-circle position under no stress, and then both pulleys are moved to a specified shaft distance. Next, misalignment is applied to one of the pulleys. Then, the drive pulley is gradually accelerated to reach a target speed, while pulley thrust pressure is being applied. In the meantime, reverse torque is gradually applied to the driven pulley.

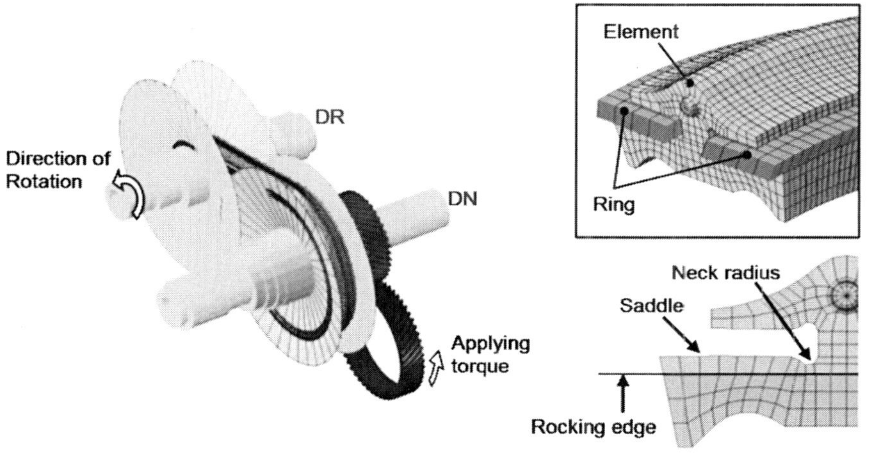

Figure 3. Metal V-belt model for predicting transmission efficiency considering shaft deformation

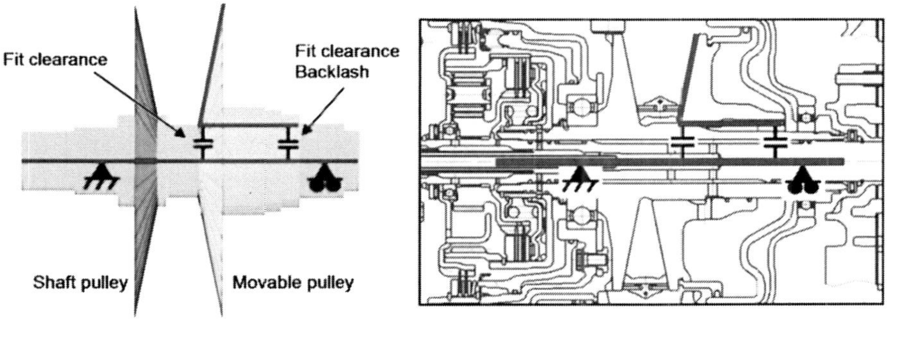

Figure 4. Boundary condition around pulley shaft

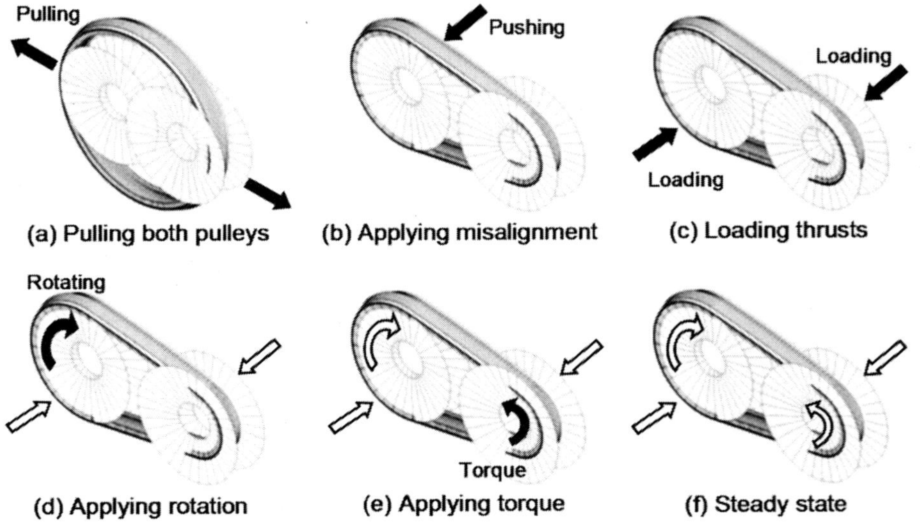

Figure 5. Analytic procedure

3.2. Belt transmission efficiency prediction using pulley thrust pressure control

The transmission efficiency η of the V-belt can be obtained by the following equation:

$$\eta = \frac{\omega dn \times Tdn}{\omega dr \times Tdr} = i\frac{Tdn}{Tdr} \tag{1}$$

ωdr : drive (DR) shaft speed

ωdn : driven (DN) shaft speed

i : ratio

Tdr : DR shaft torque

Tdn : DN shaft torque

Thus, the DR shaft torque, DN shaft torque, and ratio must be obtained to predict efficiencyη. The conventional element stress prediction model required simulations to be made based on pulley thrust pressures measured in an actual CVT. Unlike element stress measurement, the transmission efficiency can be easily measured in an actual vehicle, but it would be impractical for a simulation to require actual measurements to predict the efficiency. Accordingly, a pulley thrust controller used in the previous research [7] was implemented in the simulation to apply pulley thrust pressure that depended on each operating condition.

Figure 6 is a block diagram of the thrust controller. Enclosed in the dashed box in this figure is a traditional proportional-integral (PI) controller. This controller calculates the actual ratio from ωdr and ωdn of the belt model, and adjusts pulley thrust pressure by means of a feedback loop until the ratio reaches a target value Itarget. Controller gains were re-defined because the previous metal V-belt model was replaced with the element stress prediction model.

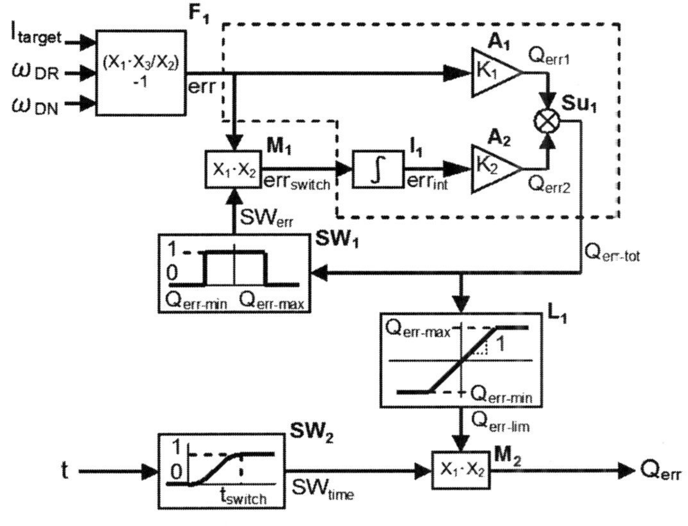

Figure 6. Thrust controller block diagram

Figure 7 shows calculations made with the metal V-belt model incorporating the above pulley thrust pressure control. In this simulation, ωdr, Tdn, and the driven pulley thrust pressure Qdn were given as input, and the drive pulley thrust pressure Qdr was controlled to make the actual ratio reach the target value. Because speed ratio error must be considered in deciding the Qdr control value, pressure was applied before starting the rotation. Then, ratio control was started after making sure that the actual ratio had been read accurately. As demonstrated in Figure 7, the speed ratio is maintained at the target value when the pulley thrust pressure is controlled by the ratio controller.

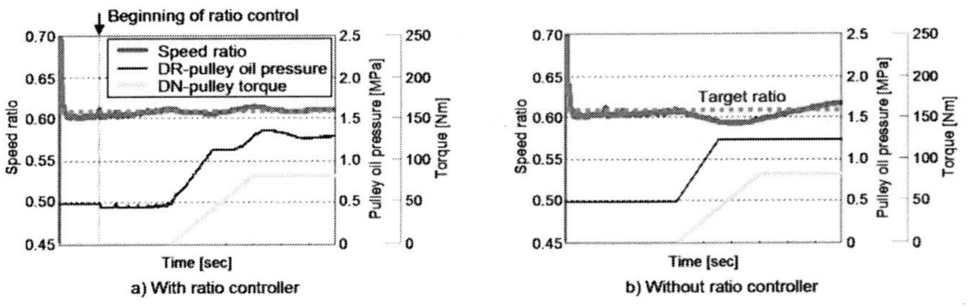

Figure 7. Effect of ratio controller

Figure 8 shows the belt transmission efficiency obtained from the computation result of this model. As shown here, the simulation implementing pulley thrust pressure control enables transmission efficiency prediction at a target speed ratio.

4. BELT TRANSMISSION EFFICIENCY AT DIFFERENT RATIO

In a CVT equipped with a metal pushing V-belt, the transmission efficiency is known to peak at the speed ratio of 1.0 (MID). The efficiency lowers gradually while a vehicle is decelerating (shifting to

LOW) or accelerating (shifting to OD). The developed simulation was used to calculate friction loss at each portion of the metal V-belt at different speed ratios.

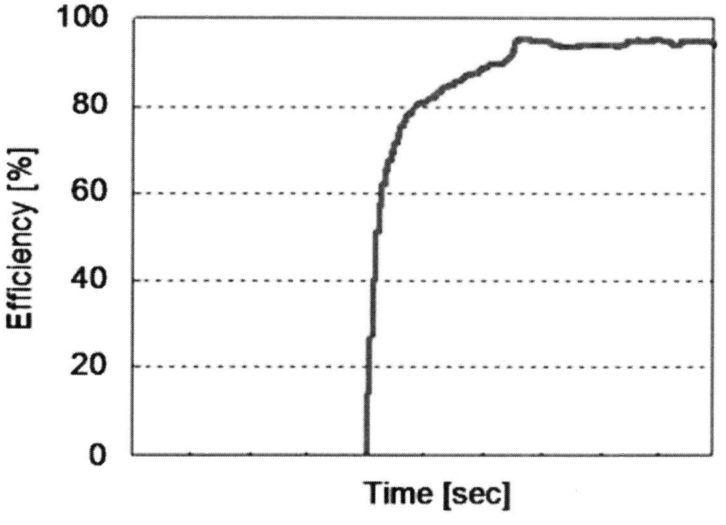

Figure 8. Calculated transmission efficiency of metal V-belt with ratio controller

4.1. Accuracy of belt efficiency prediction

Figure 9 shows belt transmission efficiency calculated for and measured at the MID, LOW, and OD ratios under a certain operating condition. Figure 10 shows the calculated and measured pulley shaft thrust load ratios. Both graphs indicate the same tendency, which verifies the validity of simulations at all ratios.

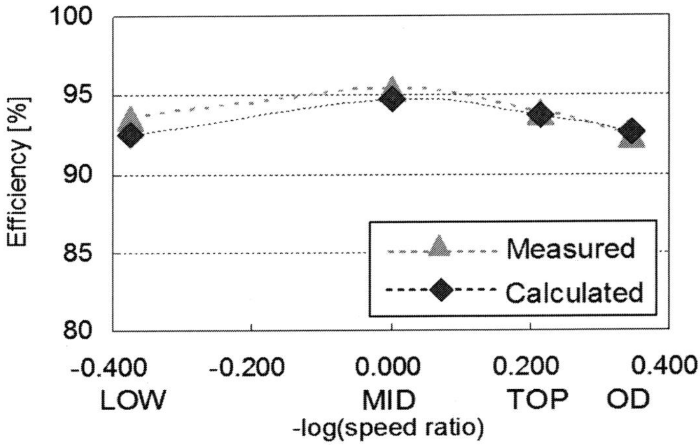

Figure 9. Belt transmission efficiency

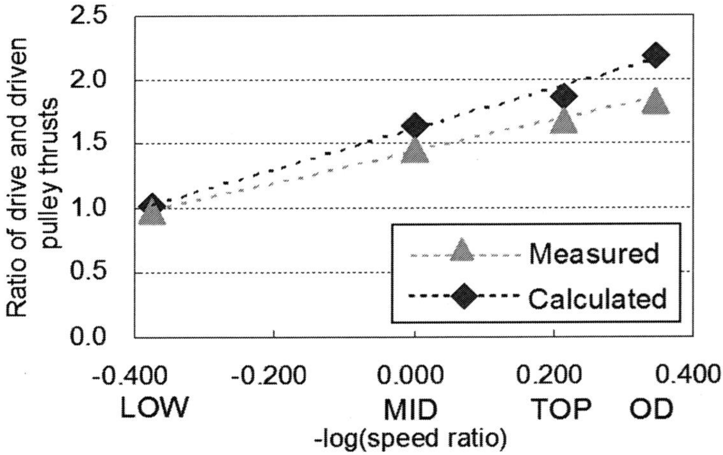

Figure 10. Ratio of drive and driven pulley thrusts

4.2. Friction loss of metal V-belt at each ratio

As mentioned before, the transmission efficiency of a metal pushing V-belt is mostly determined by friction loss that occurs between its elements and pulleys and between the elements and rings. Calculations were therefore made to compare friction losses of these areas at different ratios.

Figure 11 shows friction forces acting on the element V-surface and the element-pulley speed difference during one rotation of the belt. Similarly, Figure 12 shows friction forces acting on the element saddle surface and the element-ring speed difference. Both friction forces and speed differences are in the torque transmission direction. The speed difference is calculated for the same belt mounting position at each ratio. Figure 11 shows the speed difference between the element V-surface speed and the pulley speed, and Figure 12 the difference between the element saddle speed and the speed of the innermost ring.

When a speed difference occurs between two parts to which a friction force is applied, this speed difference may be assumed to be sliding velocity. Thus, friction loss can be obtained by multiplying the sliding velocity and the friction force. Figure 13 shows friction losses calculated in this way for the element V-surface and the saddle surface at each ratio. When attention is paid to friction loss on the element V-surface, loss is large near the drive pulley exit at the LOW ratio. In contrast, at the MID and OD ratios, friction loss is greater at the pulley entrance. On the driven pulley, loss is greater near the pulley exit, regardless of the ratio. When studied in relation to sliding velocity, friction loss becomes greater at locations where the sliding velocity is higher at all three ratios.

As for friction loss on the saddle surface, loss is smaller at the MID ratio than the other ratios. This is because, at the MID ratio, the belt mounts on a medium-diameter position on the pulley, where the sliding velocity between the pulley and the element is lower. At the other ratios, friction loss on the saddle surface is greater when it is on the smaller-diameter pulley.

A similar comparison was made with friction loss that occurs between adjacent elements and between adjacent of rings (Figure 14, Figure 15, and Figure 16).

As shown in each figure, friction loss is smaller than the loss calculated for the two aforementioned areas. This is because the sliding velocity is smaller between adjacent elements and adjacent rings where a friction force is present.

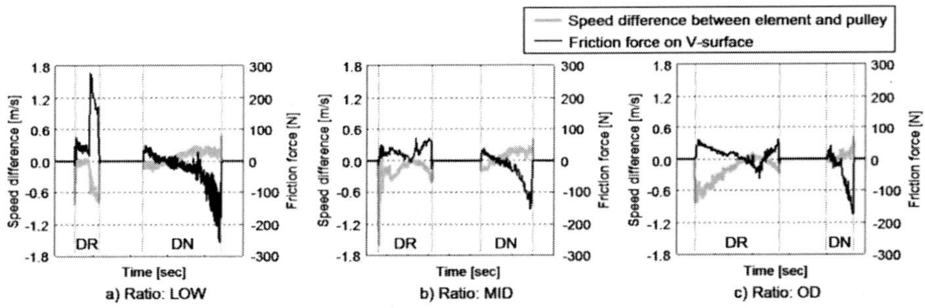

Figure 11. Element V-surface friction force and element-pulley speed difference

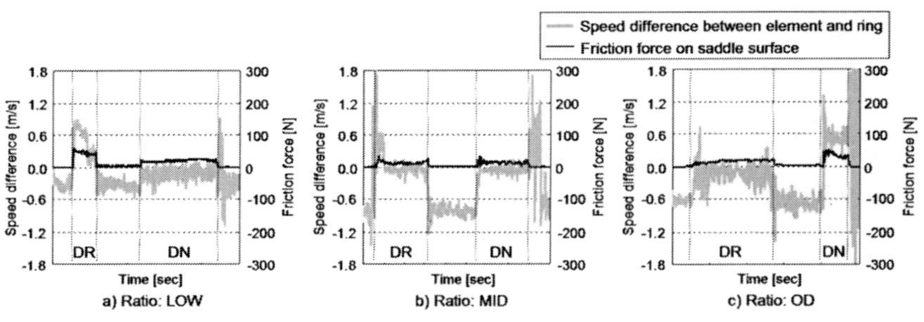

Figure 12. Element saddle-surface friction force and element-ring speed difference

Figure 17 shows the comparison of friction losses per unit time at the three speed ratios. When this figure is studied with Figure 9, it is clear that friction loss is greater at ratios with lower transmission efficiency. The comparison also reveals that the transmission efficiency is affected by friction losses on the element V surface and the saddle surface. When attention is paid to the proportion of friction losses at different ratios, friction losses on the element V-surface and the saddle surface are almost equal at the LOW ratio. At the OD ratio, friction loss on the saddle surface accounts for a dominant proportion.

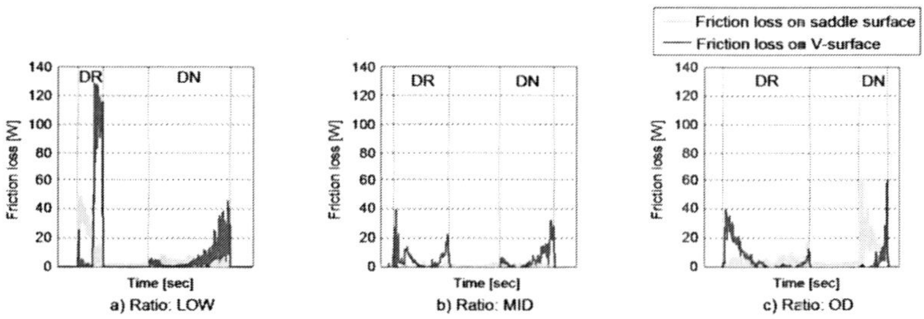

Figure 13. Friction loss on element friction surface

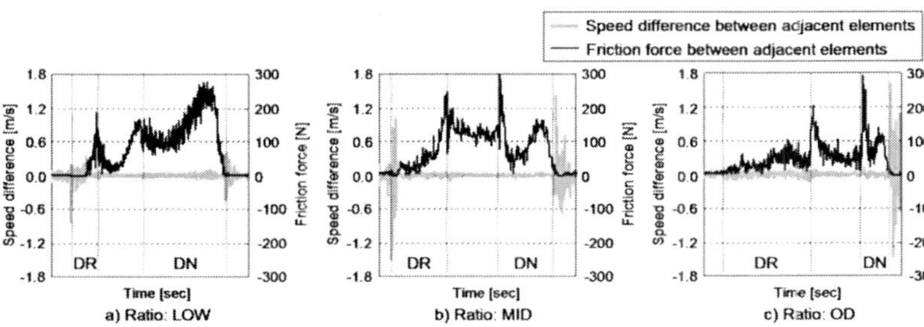

Figure 14. Friction force and speed difference between adjacent elements

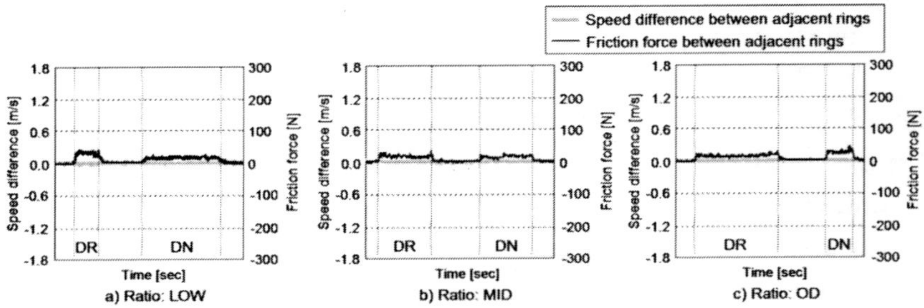

Figure 15. Friction force and speed difference between adjacent rings

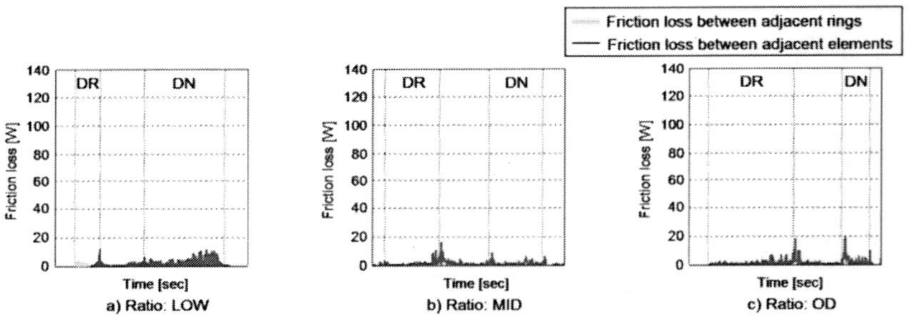

Figure 16. Friction loss between adjacent elements and between adjacent rings

5. APPLICATION FOR ACTUAL CVT CONFIGURATION

Figure 18 shows an actual CVT configuration to be calculated the transmission efficiency. Replacement of the former single-piston configuration with double pistons aligned in the axial direction has resulted in approximately 1.8 times more thrust at identical oil pressures. This has reduced line pressure when the CVT is in its frequently used overdrive ratio, thus reducing the pump workload.

In this mode, the elements and the pulleys are modeled as elastic bodies to help enable their deformation to be considered. The number of divisions in the circumferential direction was increased and a coefficient of friction was set to help enable study of the effect of the fit clearance of the fitted parts of the pulley shaft and the movable pulley and of the roller spline backlash. Figure 19 shows Von Mises stress distribution except the gears. Most of the parts are now modeled as flexible bodies.

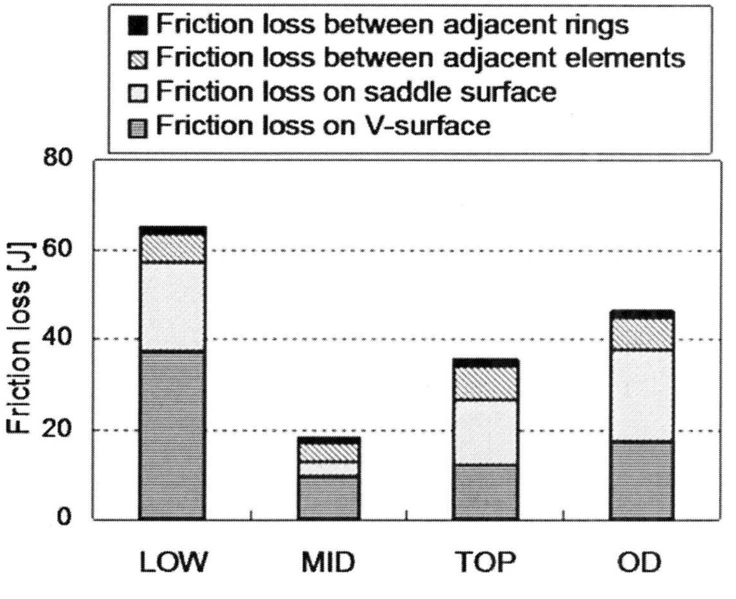

Figure 17. Belt friction loss per unit time

The transmission efficiency is now predicted using pulley thrust pressure control with appropriate control gains. (Figure 20) The speed ratio is kept at a target speed ratio during this simulation.

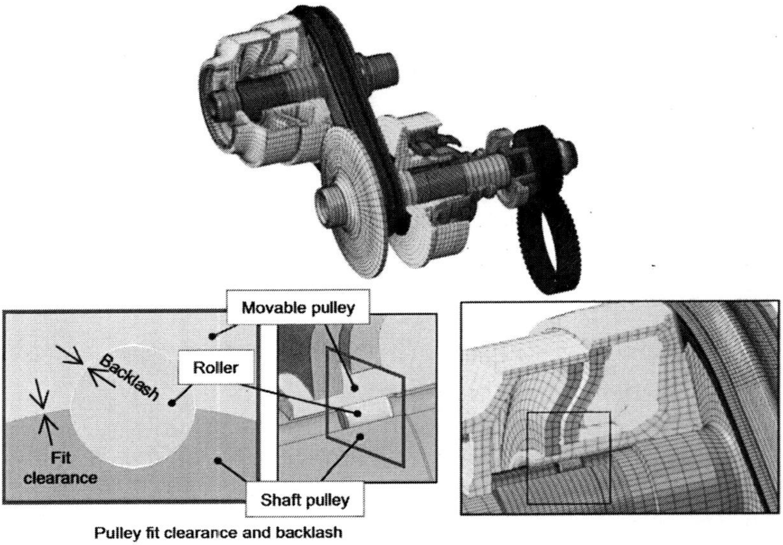

Figure 18. Metal V-belt model considering pulley V-surface deformation

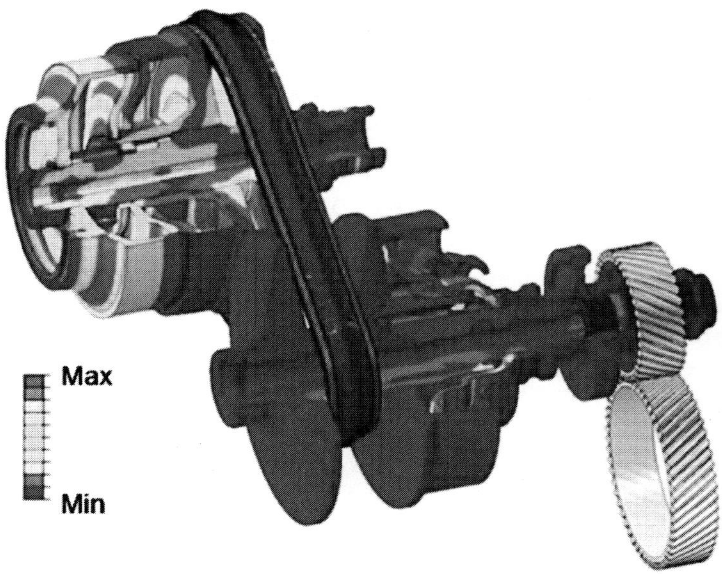

Figure 19. Stress distribution of Metal V-belt and pulleys

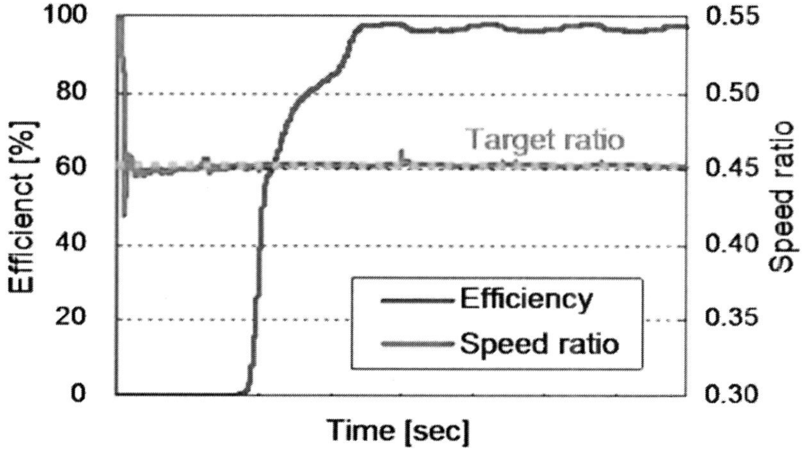

Figure 20. Calculated transmission efficiency of metal V-belt with FEM pulleys

5.1. Effect of fit clearance on transmission efficiency

In CVTs equipped with metal pushing V-belts, transmission efficiency is known to peak at the shift ratio of 1.0 (MID). Transmission efficiency declines gradually while the vehicle is decelerating (shifting to LOW) or accelerating (shifting to OD). However, the effect of the clearance of the section in which the external shape of the pulley shaft and the internal diameter of the movable pulley are in contact and of the roller on the transmission efficiency and the strength of the metal pushing V-belt is not necessarily clear. The method discussed here was therefore employed, using the OD ratio, in order to calculate friction loss for each part of the belt when the clearance of the section in which the external shape of the pulley shaft and the internal diameter of the movable pulley are in contact ("large-diameter clearance" below) and the clearance in the direction of rotation determined by the roller ("backlash" below) were varied.

Figure 21 and 22 show that when the large-diameter clearance of the fitted sections becomes narrower, friction loss on the element V-surfaces is reduced, and the transmission efficiency of the belt increases. Fig. 23 shows the changes in the winding diameter of the belt at this time. When the large-diameter clearance increases, the changes in the belt winding diameter also increase in magnitude. In other words, the belt slips in the radial direction, thus reducing the amount of friction force available for transmission, with the result that transmission efficiency declines. In addition, a comparison of the surface pressure distribution on the element V-surfaces shows that when the large-diameter clearance becomes greater, the surface pressure distribution tends towards the inside of the radius (Fig. 24). This is believed to be an effect of the fact that when the belt is transmitting torque, its effective radius is reduced, and torque transmission efficiency declines.

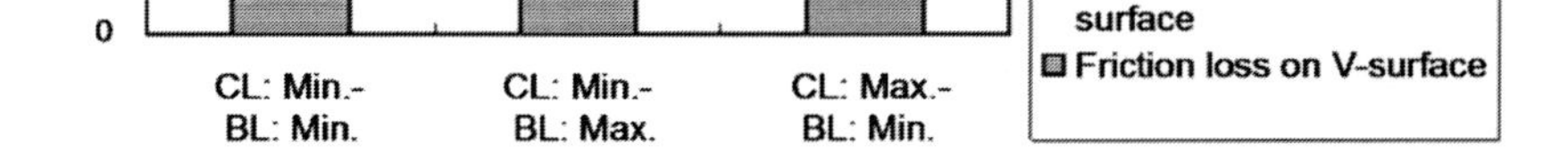

Figure 21. Fit clearance and friction loss on belt surface

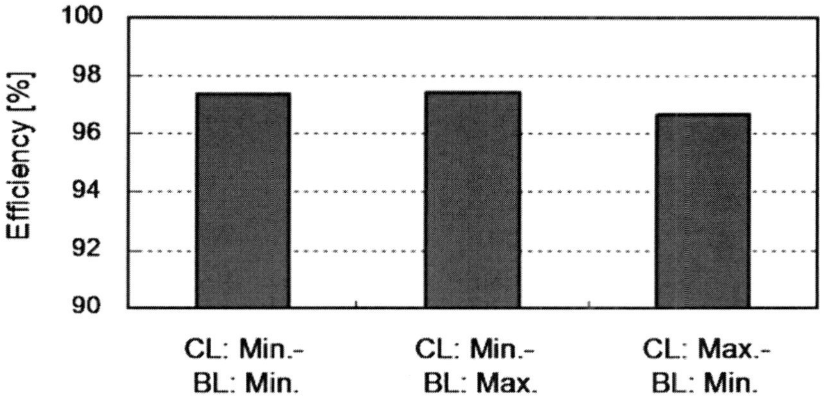

Figure 22. Fit clearance and belt transmission efficiency

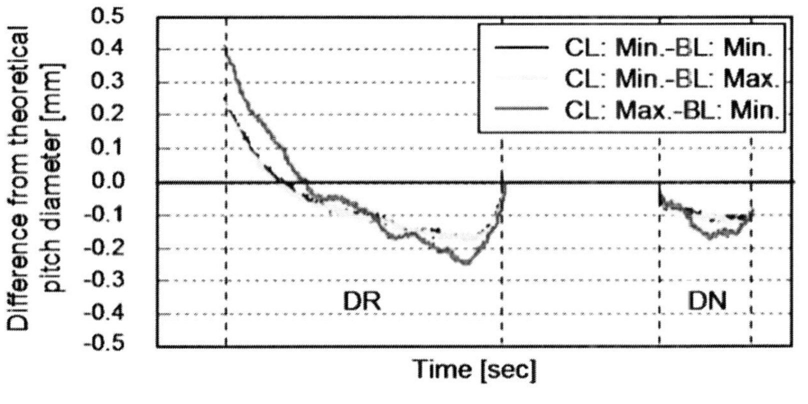

Figure 23. Fit clearance and belt pitch diameter

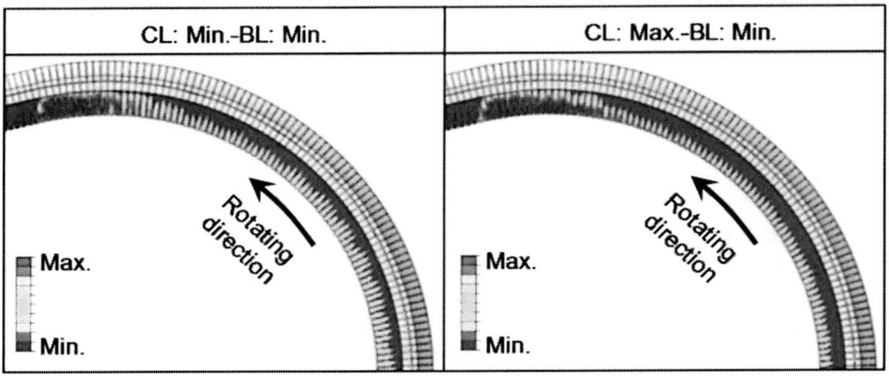

Figure 24. Fit clearance and distribution of element V-surface pressure

5.2. Effect of pulley stiffness on belt transmission efficiency

Of the parts that make up a CVT, the size and weight of the pulleys is particularly high, creating the need for the development of lighter-weight pulleys. Effects on the strength of metal pushing V-belts due to changes in pulley stiffness resulting from reduction in weight have also been reported [5,7]. As in the analyses conducted in the previous chapters, the effect on the transmission efficiency and strength of the belt when the pulley stiffness was varied was therefore considered. Altering the form of the pulleys in order to quantitatively vary stiffness would represent a challenge, and Young's modulus for the pulleys in the model discussed above was therefore altered in order to vary the stiffness. The Young's modulus of the pulleys was varied by ±20 against Young's modulus for iron, and the fit clearance was minimized.

As Figure 25 and 26 show, when the pulley stiffness is reduced, friction loss on the element V-surfaces increases, and the transmission efficiency of the belt declines. Figure 27 shows changes in the winding diameter of the belt at this time. A comparison of these results with the results of former chapters shows that the effect produced by reducing the stiffness of the pulleys displays an identical tendency to that produced by increasing the fit clearance, but the magnitude of the change is smaller.

6. DISCUSSION

These analyses focus on the steady state response under some operating conditions. However, some results show the vibration, such as speed difference between element and ring in Figure 12. These are two reasons for the vibration. One is the impact forces are generated when the element gets into and out the pulleys. Another is the each element behavior itself. Figure 28 shows the element reaction force on V-surface and the friction coefficient on V-surface. The friction coefficient in these analyses shows the vibration in the pulleys. This presents that each element could move individually to keep the contact and friction for adjacent parts.

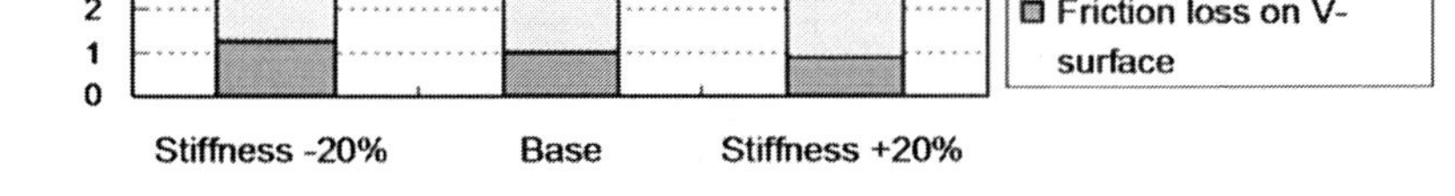

Figure 25. Pulley stiffness and friction loss on belt surface

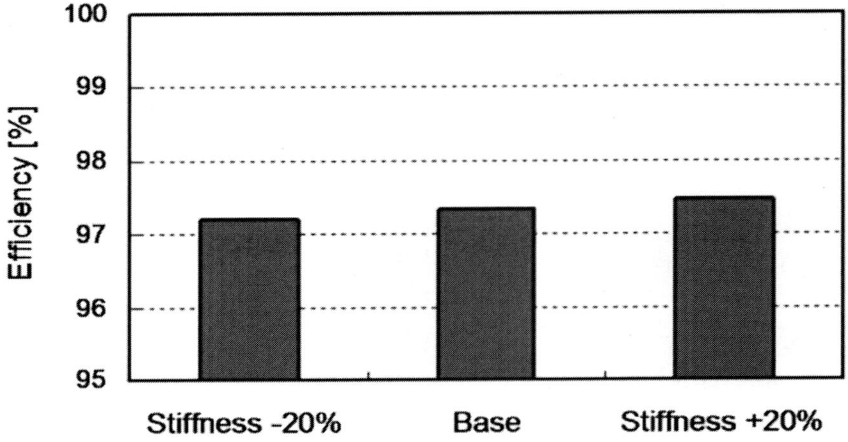

Figure 26. Pulley stiffness and belt transmission efficiency

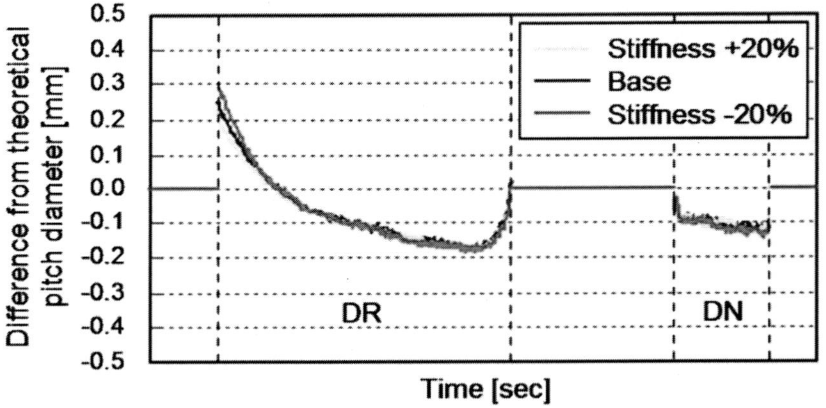

Figure 27. Pulley stiffness and belt pitch diameter

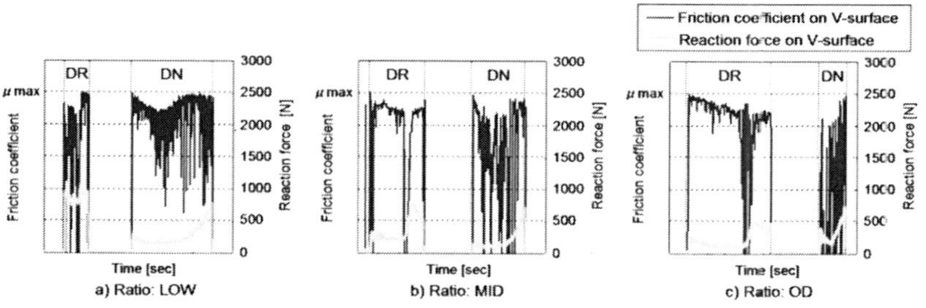

Figure 28. Friction coefficient currently used on element V-surface

7. CONCLUSION

1. Feedback control to maintain speed ratios in an actual CVT has been implemented in a metal V-belt behavior simulation, making it possible to predict the CVT transmission efficiency under an arbitrary operating condition.

2. This simulation has successfully quantified sliding velocities and friction forces of element contact areas along the entire length of the V-belt.

3. The simulation was used to calculate friction losses from sliding velocities and friction forces at different ratios—loss on the V-surface, loss on the saddle surface, loss between elements, loss between rings, and their proportions.

4. The simulation technique was also available for actual CVT configuration to predict the transmission efficiency using flexible pulley components.

REFERENCES

1. Yamamoto, K., Sakaguchi, S., Kishida, M., Kimura, E. and Abe, H. "Development of Integrated Engine-CVT Control System," Honda R&D Technical review, Vol. 11, No. 1

2. Akehurst, S., Vaughan, N. D., Parker, D. A. and Simner, D. "Modelling of loss mechanisms in a pushing metal V-belt continuously variable transmission. Part 1: torque losses due to band friction," Proc. Instn Mech. Engrs vol. 218 Part D: J. Automobile Engineering

3. Veenhuizen, P. A., Bonsen, B., Klaassen, T.W.G.L., Albers, P.H.W.M., Changenet, C. and Poncy, S. "Pushbelt CVT efficiency improvement potential of servo-electromechanical actuation slip control," 2004 CVT Congress

4. Van der Noll, E., Van der Sluis, F., Van Dongen, T. and Van der Velde. A. "Innovative Self-optimising Clamping Force Strategy for the Pushbelt CVT," SAE Paper, 2009-01-1537

5. Narita, K. and Priest, M. "Metal-metal friction characteristics and the transmission efficiency of a metal V-belt-type continuously variable transmission," Proc. IMechE vol.221 Part J: J. Engineering Tribology

6. Saito, T. "Application of Stress Simulation Under Transient Condition for Metal Pushing V-belt of CVT," SAE Paper, 2008-01-0415

7. Saito, T. and Lewis, D. A. "Development of a Simulation Technique for CVT Metal Pushing V-belt with Feedback Control," SAE Paper, 2004-01-1326

7

CHAPTER

FINITE ELEMENT ANALYSIS APPLICATIONS IN FAILURE ANALYSIS: CASE STUDIES

Ahmad Ivan Karayan[1], Homero Castaneda[1], Deni Ferdian[2, 3], Sri Harjanto[2, 3], Dwi Marta Nurjaya[2, 3] and Ahmad Ashari[2, 3]

[1] Department of Chemical and Biomolecular Engineering, The University of Akron, USA

[2] Department of Metallurgy and Materials Engineering, University of Indonesia, Indonesia

[3] Center for Materials Processing and Failure Analysis (CMPFA), University of Indonesia, Indonesia

1. INTRODUCTION

The use of the finite element in engineering applications has grown rapidly in recent years. Finite element analysis (FEA) is based on numerical computation that calculates all parameters and boundaries given. Supported with powerful computer processors and continuous software development, the finite element method is rapidly advancing. The use of finite element analysis is not limited to the engineering field, as there are also medical and geospatial applications. The early development of the finite element can be traced back to the work of Courant [1], followed by the work of Martin [2], which applied the solutions for structural analyses at Boeing Company in the 1950s. Further work by Argyris, Clough, Turner, and Zienkiewicz developed the governing mathematical equation for the finite element method [3].

The numerical simulation introduced in the 70s for the stress model on concrete as published by Hillerborg [4] is a clear example of the FEA concept. Huiskes *et al.* [5] also stated that the finite element has been used in a structural stress analysis of human bones for biomechanic applications. Zienkiewicz *et al.* [6] applied the finite element method to the linear and non-linear problems encountered during the analysis of a reactor vessel. Gallagher [7] studied brittle material design through use of the finite element method, which incorporated thermal and elastic analysis aspects of the overall design. Miller *et al.* [8] used the finite element method during the study of crack stability of a turbine blade and proposed a hypothesis based on material strength characteristics, plastic zone size/history, and residual plastic strains.

In the mid 80s, Oritz *et al.* [9] proposed a method that aimed to enhance the performance of general classes of elements that undergo strain localization. An overview of the application of the finite element in machining from the 70-90s was well documented by Mackerle [10]. The application of the method to localize fractures was studied by Broberg [11], and further work was continued by Borst *et al.* [12]. Nowadays, finite element analysis is widely used by the engineering field in fluid dynamic and electrical applications.

In general, the finite element analysis is widely used in the pre-production manufacturing process to determine the most cost-effective decision based on the analysis. Finite element simulations allow comparison between different "designs." Finite element analysis can simulate operational and environmental conditions and formulate modifications without creating a physical prototype.

In order to adequately determine the root cause of material failure, two outcomes are required: the answer and an explanation. The failure of a mechanical component is usually associated with materials, the environment, a third party, or human error. An investigation through metallurgical failure analysis is usually conducted to reveal the root cause and the failure mechanism. Work by Griffith [13] on fracture mechanics created a breakthrough in the understanding of the material fracture mechanism. In certain cases, a conventional failure analysis approach is not enough to reveal the failure, and therefore, a more comprehensive analysis through the finite element is needed. Prawoto *et al.* [14] explained that the use of the finite element is an effective approach when the causes of failure are determined using qualitative metallographic and fractographic testing. The finite element requires an understanding of how a component works and will support the correct information and data. The quality of the data provided is a key element for the successful outcome of the simulation.

2. THE FINITE ELEMENT IN FAILURE ANALYSIS

The use of finite element analysis in mechanical applications has heavily increased in recent years. The continuous efforts to improve calculations and analyses so that models accurately incorporate actual conditions have been rewarded as a consequence of computer tool evolution. The finite element analysis has become an important tool for improving the design quality in numerous applications. The finite element analysis is a computer-based technique to solve problems using numerical solutions. The analysis includes a method based on creating a geometrical model of the structure that is divided into individual nodes or elements.

Finite element modeling provides different insights into the engineering analysis that cannot be obtained with the classical failure analysis method. Classical commercial software such as ABAQUS® and ANSYS® has been widely used to analyze failures or defects and reconstruct possible root causes. The images and animation produced by the software help to give a better understanding by visualizing the root cause behind the failure event. Previously, the development of the finite element evolved slowly due to a lack of tools to solve mathematical equations, and therefore, the method remained dormant until the computer era.

The early finite element software was commissioned by NASA in mid 1960s, which introduced NASTRAN® as an application that helped to design more efficient structures for vehicles that were developed by them [15]. The ANSYS® software, which was released in 1970, was run on a massive mainframe computer that was less powerful than the personal computers of today [16]. The early software was only capable of simulating the 2D beam model, but it eventually progressed towards the modern 3D solid model of today as computer hardware developed.

The applications of finite element analysis in engineering failure analysis are in continuous evolution as more factors related to the failure event are taken into account. For example, in the case of turbine blades involved in jet engine failure, engineers should incorporate the thermal effect and load effect received by the blade in the simulation, not to mention other possibilities from foreign particles that could initiate the failure. Therefore, FEA calculations now incorporate a combination of multiple physical environments. Current research related to the application of the finite element in failure analysis has been published in peer-reviewed journals, and can be seen in publications such as *Engineering Failure Analysis Journal*® and *Failure Analysis and Prevention Journal*®.

In general, a finite element method consists of three phases: (1) pre-processing, where the analyst creates the finite element mesh and applies certain parameters or boundaries to the model, (2) solver/solution, where the program runs the governing mathematical equation that was created by the model, and (3) post-processing, where the result is evaluated and validated for further interpretation.

Some limitations must be recognized in the finite element method. The ability to define and analyze the system and the model will determine the quality of the simulation results. The evaluations of any failure event need a comprehensive approach to reveal the root cause of the failure. FEA is a tool where multidisciplinary fields are combined to help find or support the solution with better accuracy for a final conclusion. Below are three case studies that illustrate how finite element analysis can be utilized in failure analysis.

3. FEA IN MECHANICAL FAILURE

The use of FEA for mechanical failure has become a necessary tool that is easily obtained, as the commercial software is readily available for users. FEA is widely used in failure analysis assessment when analyzing and characterizing the quantitative and qualitative approaches used to determine the root cause of an event leading to a failure. The software is given historical background information related to the failure along with other boundary conditions. Below is an example of a mechanical failure analysis assessment that used finite element analysis software to verify the findings regarding the failure event.

In the first case, the finite element method was used to analyze the stress distribution of a failed 28MW horizontal hydroturbine shaft [17]. The data corresponded to the fractography and metallographic observations. The finite element analysis was performed for normal conditions as well as the type of high load conditions that would be experienced during the start-up period.

The shaft was constructed by the joining of a forged and cast part by slag welding. A crack was observed at the radius area of the casted part. The part was given an estimated 200,000 hours of operating life, and failure was detected after 163,411 hours in service. A visual examination of the fracture surface showed a fatigue pattern with obvious ratchet marks. Further observation of the fracture surface showed visible distorted fatigue lines around numerous gas holes and areas of increased porosity.

An original document indicated that the shaft was heat-treated to complete an austenization process. However, metallographic examination showed cast ferrite-pearlite with an undissolved dendrite structure, which might indicate that an improper heat treatment process occurred. In addition, a large non-metallic inclusion was also observed in the cast part.

A linear FEA was used to determine the stress state of the turbine shaft and shaft flange. The commercial ANSYS® software was used for the finite element modeling. The model represents a discrete continuum by an 8-node finite element with three degree of freedom that comprises 49,430 nodes and 47,547 elements. All of the boundary conditions were incorporated into the model.

A numerical calculation to determine the shaft flange stress states was performed for two characteristic load cases, where one load was taken from the manufacturer's documentation and the second load was the static load that occurs during start-up. Calculation of the finite element for both load conditions showed that the maximum stress was at the crack initiation site at the shaft flange.

The data collected from the chemical composition test and mechanical test showed that the material did not comply with the minimum standard required. Therefore, the crack location where the failure occurred was more susceptible to stress. Finite element analysis showed a high distribution of stress at the failed area. The finite element analysis revealed that the obtained tensile stress value on the shaft flange transition radius due to the load in case 2 was higher than recommended, and was characterized by the stress intensity factor at the crack tip being higher than the material threshold.

It was concluded that corrosion fatigue was the cause of the shaft failure [16]. The root cause for this case was improper corrosion protection at the failed area and a lack of periodical inspection, both of which were necessary due to the high stress on the region.

4. FEA IN CORROSION FAILURE

Karayan *et al.* [18] studied the failure of a seawater inlet pipe. The failure was first characterized by a small leak at approximately the 4-8 o'clock position. A schematic drawing of the inlet pipe showing the backing bar near the leak location is shown in Figure 1. This backing bar was installed on the welded surface. The visual examination of this failed pipe is shown in Figure 2. In order to find out the root cause of failure, a number of laboratory tests were performed. The results showed that the failure was caused by cavitation, as evident by the presence of a crater-like surface near the backing bar (Fig. 3). These localized craters seemed unusual since they were only noticed near the backing bar. An additional tool such as finite element analysis was used to determine why this was the case. A finite element analysis was executed based on the pipe dimension and actual fluid conditions such as velocity, pressure, temperature, and implicit parameters. Because there were no data for the initial height of the unwanted backing bar, the authors assumed that the initial height was the highest backing bar found on the specimen. Interestingly, the failure location predicted by the finite element analysis matched up with the actual evidence (Fig. 4). It precisely showed that the failure could be located around the backing bar where the eddy zone was formed in this area. This suggests that sometimes the results obtained from laboratory tests cannot explain why a failure occurs, although evidence indicates the existence of a certain problem. In this case, a finite element analysis is the only tool that can help a failure analyst find the root cause of failure. As can be seen from Figure 4, the leaks and crater-like surfaces found in the area near the backing bar were attributed to the formation of eddy zones in this area. The length of the eddy zone predicted that the area that might suffer from a flow-induced attack.

Figure 1. Big and small leak near the backing bar at about 6 o'clock position viewed from the inner side of seawater inlet pipe.

The last case is a trunkline that burst during service [19]. This incident produced a significant impact on the gas production of the company and also on the environment. The failure was characterized by a mesa-like attack and wall thinning at the 5-7 o'clock position on the inner surface. The location of failure is shown in Figure 5. A reddish brown corrosion product was noticed on the surface, but there with no indication of the occurrence of an H_2S attack found in the material. This trunkline, carrying the gas with a total pressure of 905 psi, spanned across the jungle in the descending position. The analysis of the gas composition is listed in Table 1. The topography and characteristics of the soil in which the trunkline was located and also the material specification data is listed in Table 2.

Figure 2. Big and small leak near the backing bar at about 6 o'clock position viewed from the inner side of seawater inlet pipe.

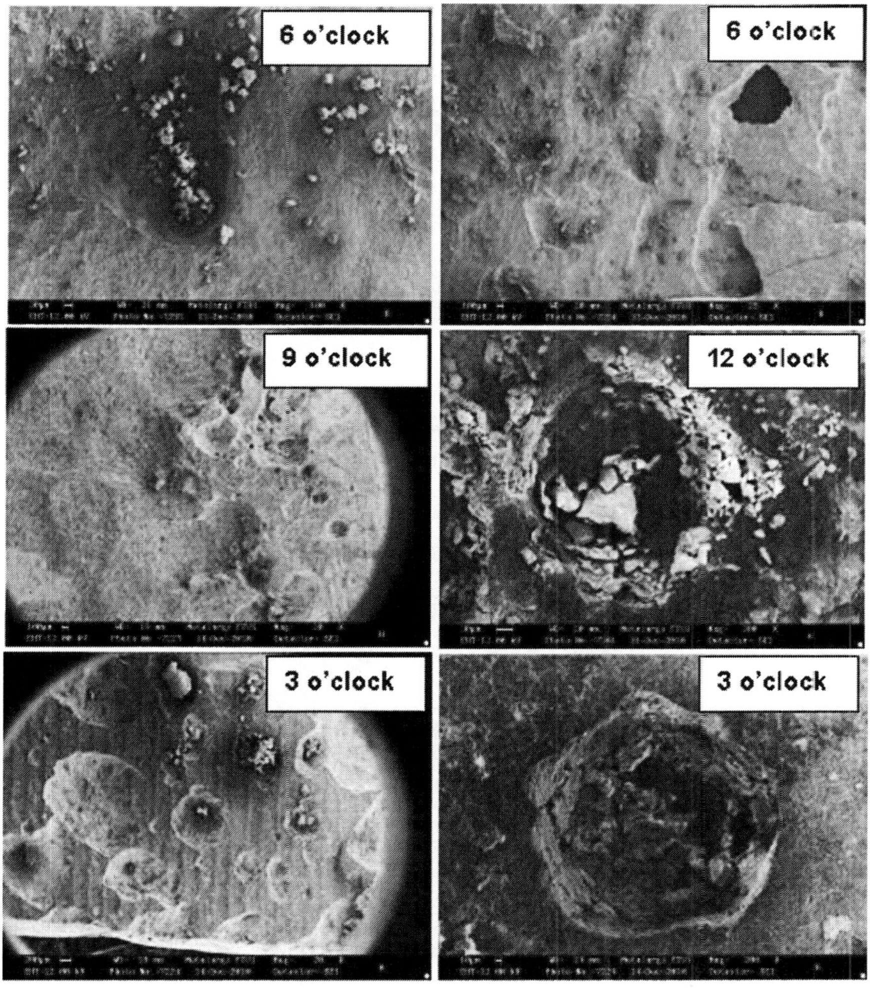

Figure 3. Surface morphologies of brown crater – like surface taken from 6 – 3 o'clock position.

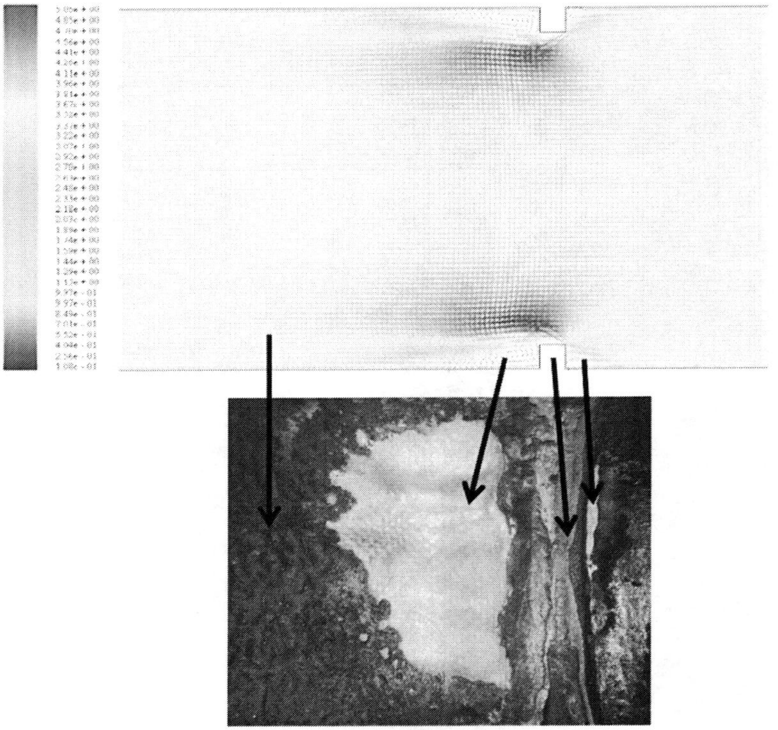

Figure 4. Finite element analysis of the inner pipe showing the orifice effect and the eddy zone near the backing bar correlated with the actual leaks on the inlet pipe.

A visual examination showed that the burst area was located at the 6 o'clock position of the (a) downstream part and the (b) upstream part (Fig. 6). Some points on the inner surface of the downstream part at the 5-7 o'clock position showed surface degradation with wall thinning and pits (Fig. 7). Some points on the inner surface of the upstream part at the 5-7 o'clock position also showed surface degradation with wall thinning and pits (Fig. 8). A uniform attack at the burst area from the 5-7 o'clock position is shown in Figure 9.

Figure 5. The specimen sent to the laboratory for analysis is indicated by arrow

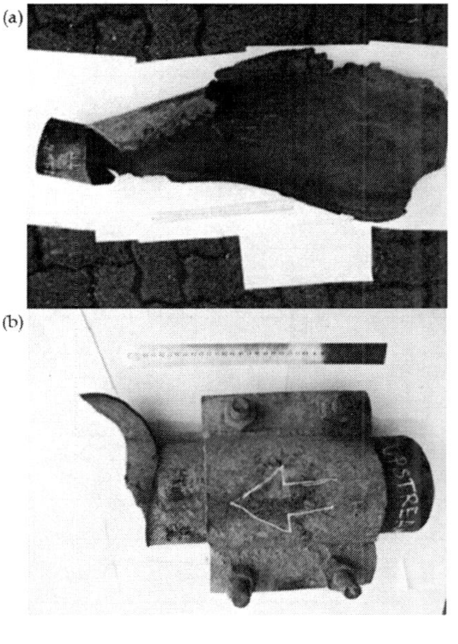

Figure 6. a) Downstream part and (b) upstream part showing the burst located at 6 o'clock position.

Table 1. esult of gas analysis in trunkline

Component		Mole %
H_2	Hydrogen	0.0000
O_2	Oxygen	0.0084
N_2	Nitrogen	3.6036
CO_2	Carbon Dioxide	1.9406
H_2S	Hydrogen Sulfide	0.0000
C_1	Methane	82.7838
C_2	Ethane	6.7222
C_3	Propane	3.0077
iC_4	Iso-Butane	0.6144
nC_4	n-Butane	0.6610
iC_5	Iso-Pentane	0.2384
C_5	n-Pentane	0.1418
C_6	Hexanes	0.1352
C_7	Heptanes	0.0982
C_8	Octanes	0.0306
C_9	Nonanes	0.0072
C_{10}	Decanes	0.0069
C_{11+}	Undecanes plus	0.0000

Table 2. Fluid properties and failed trunkline condition.

Item	Description	Result
Property	Q	± 3 MM
	WC	-
	GOR	-
	P (Psi)	750
	T (F)	140
Fluid Composition	H_2S	0.00
	CO_2 (%)	1.9406
	SRB content	-
	Chloride (%)	0.7635
	Water (%)	0.0898
Pipe Arrangement	Laydown/buried/support w/ith trestle	Laydown
	Seam position	-
Corrosion Form	Uniform/localized/pitting/etc	pitting/localized
	external/internal/both	Internal
	On seam/not	-
	On 6 o'clock/others	6 o'clock
Environment	Elevation	Descending
	Any river/road crossing	No
	Soil pH	-
	Any trees/bushes	Yes
Document	Work over on wells using the line	No job in last one year
	WT inspection	not yet

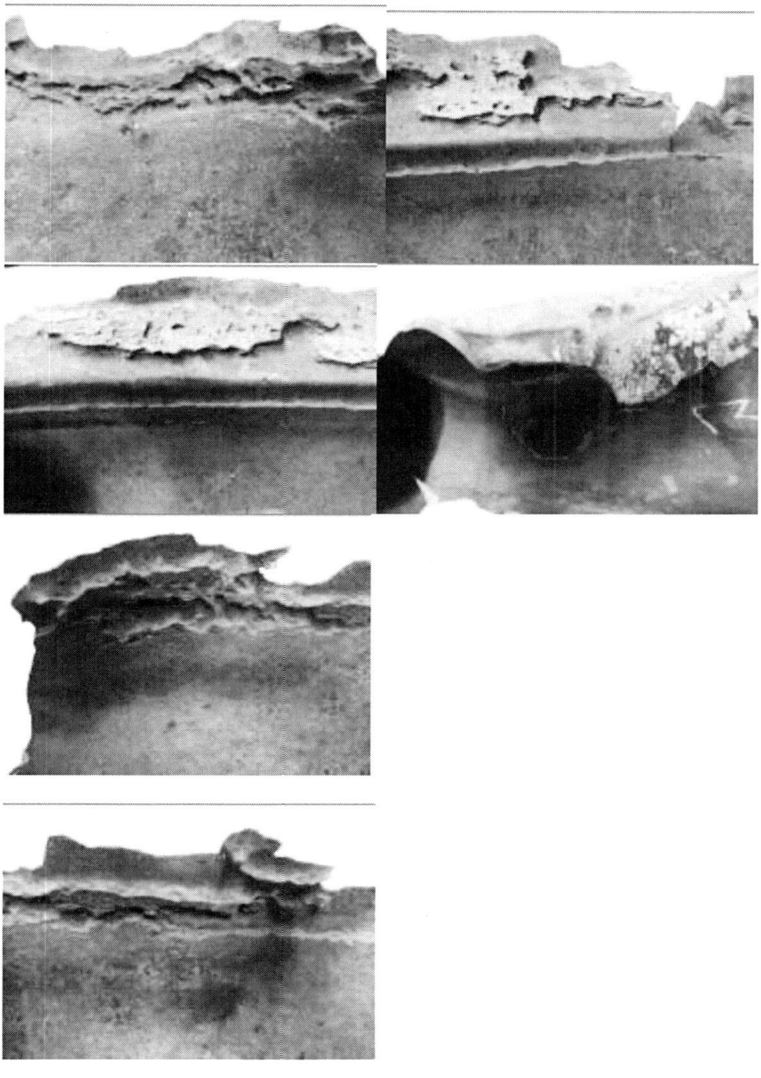

Figure 7. Wall thinning and pits at 5 – 7 o'clock position of downstream part.

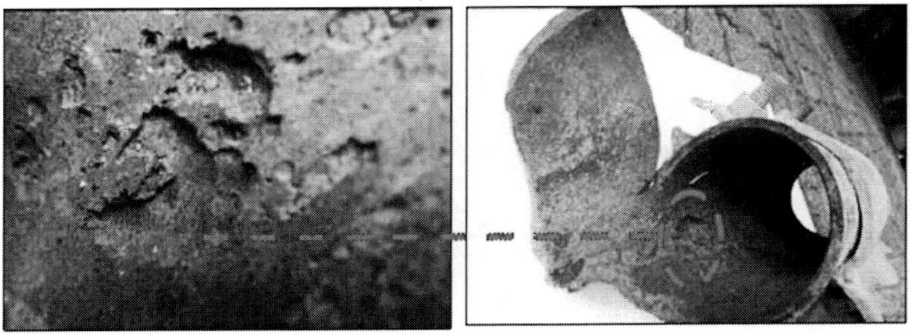

Figure 8. Wall thinning and pits at 5 – 7 o'clock position of upstream part.

Figure 9. Uniform attack at the burst area from 5-7 o'clock position.

The metallographic preparation and macroetching were performed on the perimeter of the trunkline, and the results showed that the trunkline was made of seamless pipe (Figs. 10 and 11). The thinning area shown in Figure 9 was an ERW (electric resistance welding)-free area (Fig. 10). This information indicates that the failure could not be attributed to the ERW pipe issue.

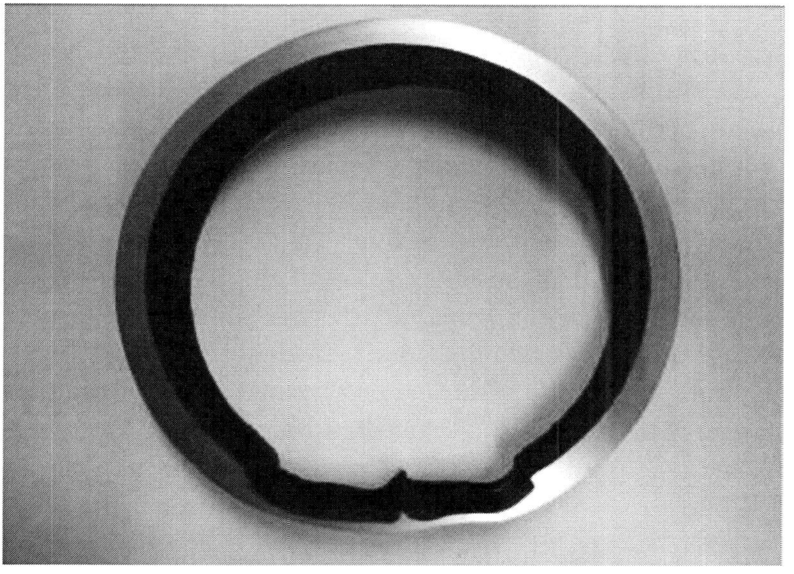

Figure 10. Results of macroetching showing the absence of ERW.

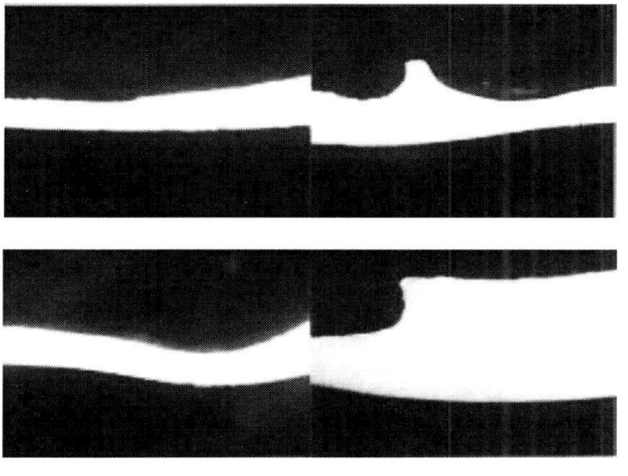

Figure 11. Macrograph of 5-7 o'clock position showing the absence of ERW.

The microstructure of the trunkline was taken from the cross-section, and the results are displayed in Figure 12. As can be seen, the trunkline is composed of ferrite (light phase) and pearlite (dark phase) with equiaxed grains, which is typical of seamless pipe. The chemical composition of the trunkline was tested using an optical emission spectrometer and the results (Table 3) show that this trunkline is composed of an API 5L X60 steel [20]. Due to the insufficient geometry of the trunkline, the mechanical property of the trunkline was only examined by hardness testing. In order to verify that the specification of this material was API 5L, the resultant hardness values were converted to tensile strength (Table 4).

Figure 12. Microstructure of trunkline showing a typical seamless microstructure with equiaxed grains. The dark phase is pearlite and the light one is ferrite.

Table 3. Chemical composition of trunkline

Material	C (%)	Mn (%)	P (%)	S (%)	Ti (%)
API 5L X60 PSL 1 [20]	0.22	1.40	0.030 (max) <	0.030 (max)	0.040
Failed Trunkline	0.165	0.594	0.003	< 0.003	< 0.002

Table 4. Converted tensile strength in comparison with API 5LX60 specification.

Material	*Average HRB*	*Approxmate UTS Based on Conversion (MPa)*
Failed trunkline	82	524
API 5L X60 [20]	-	517 (min)

The finite element analysis was executed around the overfill, and the results (Fig. 13) showed that the area at which the pipe burst had a dead or eddy zone due to the excessive overfill (Figs. 14 and 15). Because there was no information about the initial overfill height, we assumed an overfill height of 1 cm for the finite element simulation. This selection was based on the fact that there was one point at the remaining overfill that had a height of 1 cm. The other areas were mostly degraded, with heights of less than 0.5 cm. Severe corrosion was noticed on every peak of overfill.

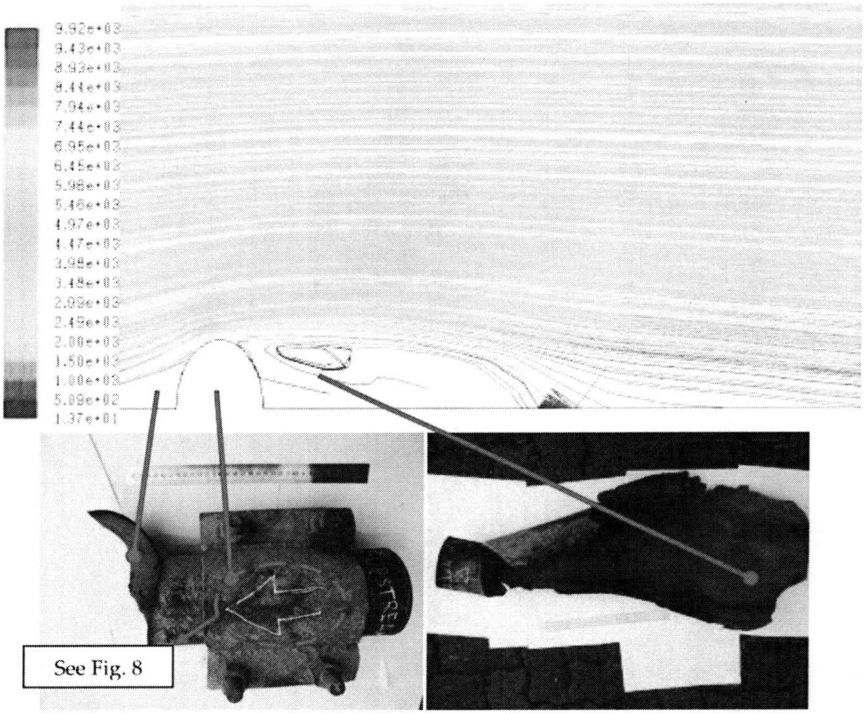

Figure 13. Finite element analysis showing the eddy zone due to overfill in comparison with the actual failed area of pipe.

As can be seen in Table 1, the only corrosive gas in the fluids flowing in this failed trunkline was CO_2. Carbon dioxide systems are one of the most common environments in the oil and gas field industry where corrosion occurs. In a relatively slow reaction, carbon dioxide forms a weak acid known as carbonic acid (H_2CO_3) in water, but the corrosion rate of CO_2 is greater than that of carbonic acid. Cathodic depolarization may occur, and other attack mechanisms may also occur. The presence of salts is relatively unimportant in sweet (CO_2) service [21], and thus, the presence of chloride in this system (0.7635%) did not significantly contribute to the failure.

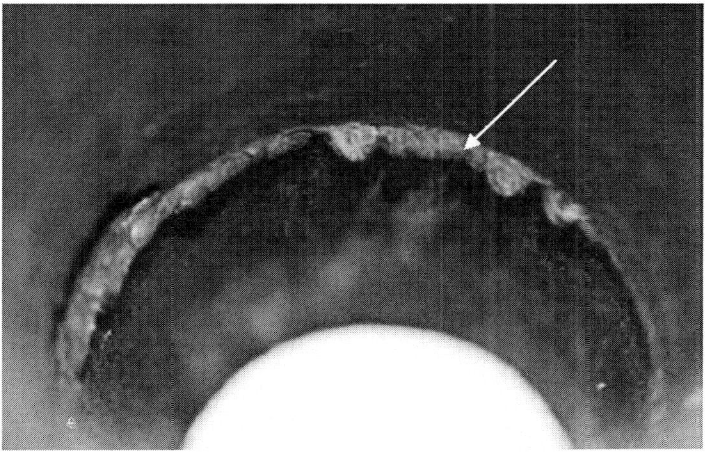

Figure 14. The inner side of circle area in Fig. 11 showing an excessive overfill (arrow) in the failed pipe at 12 o'clock position.

Figure 15. An excessive overfill and severe surface degradation beside overfill at 6 o'clock position.

The sweet environment of this system with a CO_2 partial pressure of 17.56 psig may influence the presence of corrosion. The relationship between corrosion tendency and CO_2 partial pressure in the sweet environment with pH 7 or less has been reported elsewhere [22]. When the CO_2 partial pressure is less than 7 psig, the system is non-corrosive. When it is somewhere between 7-30 psig, corrosion in the system may be present. Lastly, when it is higher than 30 psig, the system is corrosive [22]. Temperature and flow regime are closely linked because CO_2 corrosion is dynamic and very sensitive to electrochemical and physical imbalances (especially fluctuating pressure, temperature, and volume). Steady-state (P,T,V,) conditions tend to promote protective film compaction, and therefore, passivation and low corrosion rates. Lower temperatures <120°F (approximately 50°C) tend to promote patchy corrosion with softer multi-layered iron carbonate (siderite) scales that provide some barrier protection up to 140-160°F (60-70°C). Above these temperatures, damaging localized corrosion is observed as films lose stability and spall off, giving rise to galvanic mesa attack [23]. The failed pipe we studied with an operating temperature of 140°F might have formed a protective film. However, the phenomenon of film removal and its effect on the failure of this trunkline was not evident in our laboratory test. Our finite element analysis in the failed area in Figure 13 shows the unstable and chaotic flow in the failed area (called the dead or eddy zone). As illustrated, the eddy zone was triggered by the excessive overfill. This suggests that the pipe surface in this eddy area was severely attacked by flow. The presence of a passive carbonate layer could not protect the surface from this type of flow-induced attack that led to a mesa attack.

5. CONCLUSIONS

The three case studies discussed in this chapter have clearly shown us that finite element analysis (FEA) is an excellent and powerful tool that can be employed in failure analysis. Finite element analysis provides a failure analyst with more quantitative and qualitative information about the causes of failure. Although visual examinations and laboratory tests may not be able to determine a failure mechanism, the results of finite element analysis will support all data obtained from these tests. As long as a competent analyst running the finite element analysis is given sufficient data and has good knowledge of the system under study, the results of FEA will be reliable, although they should always be validated with experimental or real condition information.

REFERENCES

1. F. Williamson, Richard Courant and the finite element method: A further look, Historia Mathematica, Vol 7(4), 1980, pp. 369-378.

2. H. C. Martin, Large Deflection and Stability Analysis by Direct Stiffness Method, JPL Technical Report No. 32-931, California Institute of Technology, August 1966.

3. R. W. Clough, Early History Of The Finite Element Method From The View Point of a Pioneer, International Journal For Numerical Methods In Engineering, 60, 2004, pp. 283-287.

4. A. Hillerborg, M. Modéer, P.-E. Petersson, Analysis of crack formation and crack growth in concrete by means of fracture mechanics and finite elements, Cement and Concrete Research,Vol 6(6), 1976, pp. 773-781.

5. R. Huiskes, E.Y.S. Chao, A survey of finite element analysis in orthopedic biomechanics: The first decade, Journal of Biomechanics, Vol 16(6) 1983, pp. 385-409.

6. O.C. Zienkiewicz, D.R.J. Owen, D.V. Phillips, G.C. Nayak, Finite element methods in the analysis of reactor vessels, Nuclear Engineering and Design, 20(2)1972507541

7. R.H. Gallagher, Finite element analysis in brittle material design, Journal of the Franklin Institute, 290(6)1970523537

8. R.E. Miller Jr., B.F. Backman, H.B. Hansteen, C.M. Lewis, R.A. Samuel, S.R. Varanasi, Recent advances in computerized aerospace structural analysis and design Computers & Structures, 7(2)1977315326

9. M.Ortiz, Y.Leroy, A.Needleman, A finite element method for localized failure analysis, Computer Methods in Applied Mechanics and Engineering, Vol 61(2) 1987, pp. 189-214.

10. J. Mackerle, Finite-element analysis and simulation of machining: a bibliography (1976-1996), Journal of Materials Processing Technology, Vol 86, Issues 1-3, 1998, pp. 17-44.

11. K.B. Broberg, The foundations of fracture mechanics, Engineering Fracture Mechanics,Vol16(4)1982497515

12. R. Borst, De , L. J. Sluys, H. Muhlhaus, B. Pamin, J. , Fundamental Issues In Finite Element Analyses Of Localization Of Deformation Engineering Computations, Int J for Computer-Aided Engineering, 10199399121

13. Griffith: Phil Trans Roy Soc1921221163198

14. Y.Prawoto, Quantitative failure analysis using a simple finite element approach.Journal of Failure Analysis and Prevention, 10 (1) 2010810

15. http://en.wikipedia.org/wiki/Nastran

16. http://www.odonnellconsulting.com/forensicfea.html

17. D. Momcilovic, Z. Odanovic, R.Mitrovic, I.Atanasovska, T.Vuherer, Failure analysis of hydraulic turbine shaft, Engineering Failure Analysis, 20 (2012) pp. 54-66.

18. A.I. Karayan, A. Hersuni, D. Adisty, A. Yatim, Failure analysis of seawater inlet pipe, Journal of Failure analysis and Prevention, 11(2011) pp. 481-486.

19. A.I. Karayan, Failure Analysis of Trunkline: An Internal Report, Center for Materials Processing and Failure Analysis (CMPFA), 2011.

20. American Petroleum Institute (API)L Standard.

21. http://octane.nmt.edu/WaterQuality/corrosion/CO2.aspx

22. Garverick L, Corrosion in the petrochemical industry, pp. 92, ASM International, 1994.

23. B. Sing, K. Krishnathasan, Pragmatic effects of flow on corrosion prediction, NACE corrosion confeence and expo 2009Paper 09275

8

CHAPTER

FINITE ELEMENT METHOD APPLIED TO THE EIGENVALUE ANALYSIS OF FLEXIBLE ROTORS SUPPORTED BY JOURNAL BEARINGS

Wéderley M. Miranda[1], Marco Tulio C. Faria[2]

[1]Centro Universitário UNA, Belo Horizonte, Brazil

[2]Universidade Federal de Minas Gerais, Belo Horizonte, Brazil

ABSTRACT

This work deals with a finite element procedure developed to perform the eigenvalue analysis of damped gyroscopic systems, represented by flexible rotors supported on fluid film journal bearings. The rotor finite element model is based on the Timoshenko beam theory, accounting for the shaft rotary inertia and gyroscopic moments. The governing equations for the hydrodynamic journal bearing are obtained through the Galerkin weighted residual method applied to the classical Reynolds equation. A perturbation scheme on the fluid film governing equation permits to obtain the zero-th and first order lubrication equations for the bearings, which allow the computation of the dynamic force coefficients associated with the bearing stiffness and damping. The rotor-bearing system equation, which consists of a case of damped gyroscopic equation, is rewritten on state form to compute the complex eigenvalues. The natural frequencies at several operating conditions are obtained and compared to the technical literature data. The influence of the effective damping on the eigenvalue real part sign is analyzed for some examples of rotor-bearing systems, showing how the stability conditions can be predicted by the eigenvalue analysis. The procedure implemented in this work can provide useful guidelines and technical data about the selection of the more appropriate set of bearing parameters for rotating machines operating at stringent conditions.

Keywords: Flexible Rotors; Fluid Film Bearings; Rotor-Bearing Systems; Finite Element; Eigenvalue Problem

1. INTRODUCTION

Researchers have been continuously devising experimental, analytical and computational procedures to analyze the several dynamic aspects associated with rotating shafts employed on high-speed machines. Since 1970, the finite element method has been largely used to develop models for flexible rotors and to perform analyses of balancing, stability and torsional vibration of rotating machinery [1] -[7] . For a rotating shaft, the Timoshenko beam theory has been employed to build finite element models very accurately to analyze the dynamics of flexible rotors [5] .

Computational procedures are able to predict the dynamic response of high-speed rotors supported on fluid film bearings have been the goal of many turbomachinery manufacturers [8] . Those procedures are very useful at the preliminary design stages and commissioning of industrial rotating machines employed on the oil industry and petrochemical plants [9] . The eigenvalue analysis of rotating and stationary components of machines and mechanical equipments has been a basic step in any dynamic analysis of rotating systems [10] . Vibration modes associated with the rotating shaft and bearing support have provided important subsides for the development of computational procedures on vibration analysis, balancing techniques and monitoring of high speed rotating machinery [8] [11] -[13] .

On the eigenvalue analysis of structural dynamic systems, the governing equations are generally based on both the undamped and damped gyroscopic systems [14] . On the other hand, for industrial turbomachinery and rotating machines, the eigenvalue analysis has been carried out based on both the nongyroscopic and gyroscopic systems, taking into account or not the dissipative properties [10] [15] -[18] . The eigenvalue problem is also important on the sensitivity analysis of the system dynamic response [19] -[22] . The stability analysis and the dynamic response of gyroscopic systems can also be performed from the eigenvalue problem [17] [23] -[25] .

This work deals with a finite element procedure devised to perform the eigenvalue analysis of high-speed rotating machines supported on fluid film journal bearings. The Timoshenko beam theory is applied on the rotating shaft finite element modelling, accounting for the shear effects, the gyroscopic moments and the rotatory inertia. Lumped masses are used to model mechanical components rigidly attached to the rotating shaft, which may represent any rotating part of a turbomachine shaft, such as turbine wheels, compressor disks or pump impellers. The hydrodynamic journal bearing finite element modelling is based on the classical Reynolds equation. A linearized perturbation method is applied on the Reynolds equation to render the lubrication equations capable of predicting the eight linearized dynamic force coefficients associated with the bearing stiffness and damping. The rotor-bearing system equation, which consists of a case of damped gyroscopic equation, is rewritten on state form [26] - [28] to compute the complex eigenvalues. The natural frequencies are obtained for rotating machines operating at stringent conditions. The influence of the effective damping on the eigenvalue real part sign is analyzed for some examples of high-speed rotor-bearing systems. Also, the influence of the bearing damping coefficients on the natural frequencies and on the stability of flexible rotors is shown through some curves presented in this work. The effective damping of rotor-bearing systems is demonstrated to be a very important design parameter for high-speed rotating machinery.

2. FINITE ELEMENT EQUATIONS

The rotor-bearing system is modelled using finite element models for both the flexible shaft and the hydrodynamic journal bearings. A global equation of motion, Equation (1a), is obtained from the finite element matrices, where [M] represents the global shaft translational inertia matrix, [N] represents the global rotatory inertia matrix, [K] the shaft and bearing stiffness matrix and [C] is the generalized shaft and bearing damping matrix, in which the shaft gyroscopic effects are included. The bearings stiffness $[K_m]$ and damping $[C_m]$ coefficients are included into the system matrices, in order to represent the fluid film resistance to the rotor displacement and to velocity, respectively. The rotor-bearing system equation is rewritten on state form to compute the complex eigenvalues. The complex eigenvalues associated with the system are separated to get the natural frequencies and information on the stability of the rotor-bearing system.

2.1. Shaft Modelling

The finite element method is applied for the modelling of both the flexible shaft and the hydrodynamic journal bearings. Figure 1 depicts a schematic view of a flexible rotor supported on fluid film plain cylindrical journal bearings.

The finite element shaft modelling implemented in this work has been based on the special shape functions derived by [5] . Reference [5] employs the Timoshenko beam theory to derive the governing equations for a flexible circular shaft supported on elastic supports taking into account the shaft shear effects, gyroscopic moments and rotatory inertia. The system is represented schematically in Figure 1.

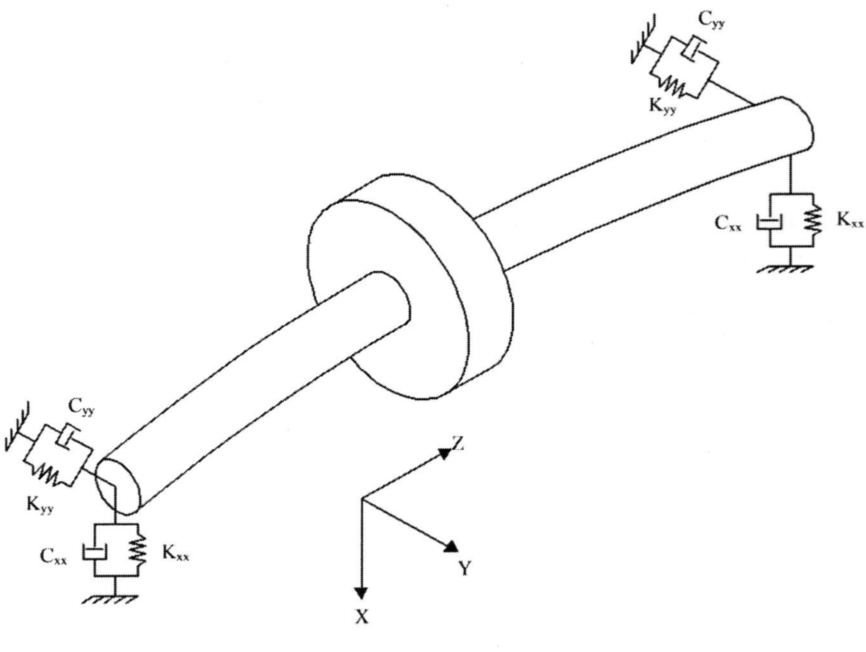

Figure 1. Flexible shaft supported on fluid film journal bearings.

Two node beam finite elements with eight degrees-of-freedom are employed to model the lateral motion of flexible shafts. The journal bearing contributions to the rotor stiffness and damping coefficients are accounted for. The finite element procedure is based on the following global equation of motion

$$[M+N]\{\ddot{U}\}+[C]\{\dot{U}\}+[K]\{U\}=\{R\} \tag{1a}$$

where [M] represents the global shaft translational inertia matrix, [N] represents the global rotatory inertia matrix, [K] the shaft and bearing stiffness matrix and [C] is the generalized shaft and bearing damping matrix, which is expressed as $[C]=[C_1]-\Omega\cdot[G]$, in which [G] is the shaft gyroscopic effect matrix. The matrix $[C_1]$ represents the bearing damping. The shaft acceleration, velocity and displacement vectors are given, respectively, by $\{\ddot{U}\},\{\dot{U}\},\{U\}$, and Ω is the shaft rotating speed (rad/s). Each node has 4 degrees of freedom, where the i-th element displacement and rotation is represented by the vector U_i composed of the following components:

$$U_i=\begin{Bmatrix} y_i \\ x_i \\ \varphi_i \\ \theta_i \end{Bmatrix}, \text{ where, } \begin{matrix} y_i = \text{horizontal displacement} \\ x_i = \text{vertical displacement} \\ \varphi_i = \text{rotation around the } y \text{ axis} \\ \theta_i = \text{rotation around the } x \text{ axis} \end{matrix} \tag{1b}$$

The external excitation force is represented by the vector {R} in Equation (1a). For the eigenvalue problem analysis, this vector is null {R} = {0}.

2.2. Bearing Modelling

The journal bearing finite element model is developed based on the classical Reynolds equation for oil-lubricated plain cylindrical journal bearings [1] . For the coordinates (X, Z), this equation is given by.

$$\frac{\partial}{\partial X}\left(\frac{h^3}{12\mu}\frac{\partial P}{\partial X}\right)+\frac{\partial}{\partial Z}\left(\frac{h^3}{12\mu}\frac{\partial P}{\partial Z}\right)=\frac{\Omega R}{2}\frac{\mathrm{d}h}{\mathrm{d}X}+\frac{\partial h}{\partial t} \tag{2}$$

The journal rotational speed is denoted by W. Journal eccentricities on the vertical and horizontal directions are expressed as e_X and e_Y, respectively. The eccentricity ratio is defined as e = e/c, where $e^2 = e_X^2 + e_Y^2$. The circumferential coordinate X = R·q and R is the bearing radius. Fluid viscosity is given by m, P represents the hydrodynamic pressure and h is the fluid film thickness. A linearized perturbation procedure is used in conjunction with Equation (2) to render the zerothand first-order lubrication equations [29] . These equations allow the computation of the bearing reaction forces and eight dynamic force coefficients. For brevity, these equations and the validation of the finite element procedure for the bearing dynamic coefficients are omitted in this work.

The dynamic force coefficients are represented in matrix form by the stiffness $[K_m]$ and the damping $[C_m]$ matrices as in Equation (3), given by [30] . They stand for the fluid film resistance to the rotor displacement and velocity, respectively.

$$[K_m]=\begin{bmatrix} K_{XX} & K_{XY} \\ K_{YX} & K_{YY} \end{bmatrix};\ [C_m]=\begin{bmatrix} C_{XX} & C_{XY} \\ C_{YX} & C_{YY} \end{bmatrix} \tag{3}$$

Figure 2 depicts the cross-section of a journal bearing and its linearized stiffness and damping coefficients along the X-axis and Y-axis.

2.3. Eigenvalue Problem

The vibration analysis of rotor-bearing systems can be carried out through computational procedures developed specially to predict the dynamic response and stability analysis of rotating shafts supported by fluid film bearings. At the preliminary design and commissioning stages of industrial turbomachinery, those procedures can bring important insights on the rotating system dynamic behaviour.

The first step in the dynamic analysis consists of obtaining the system natural frequencies under several operating conditions. The free vibration problem associated with linear systems of differential equations leads naturally to the eigenvalue problem [31] . For damped gyroscopic systems, the complex eigenvalues and eigenvectors provide very useful data about the mode shapes and stability of rotating systems.

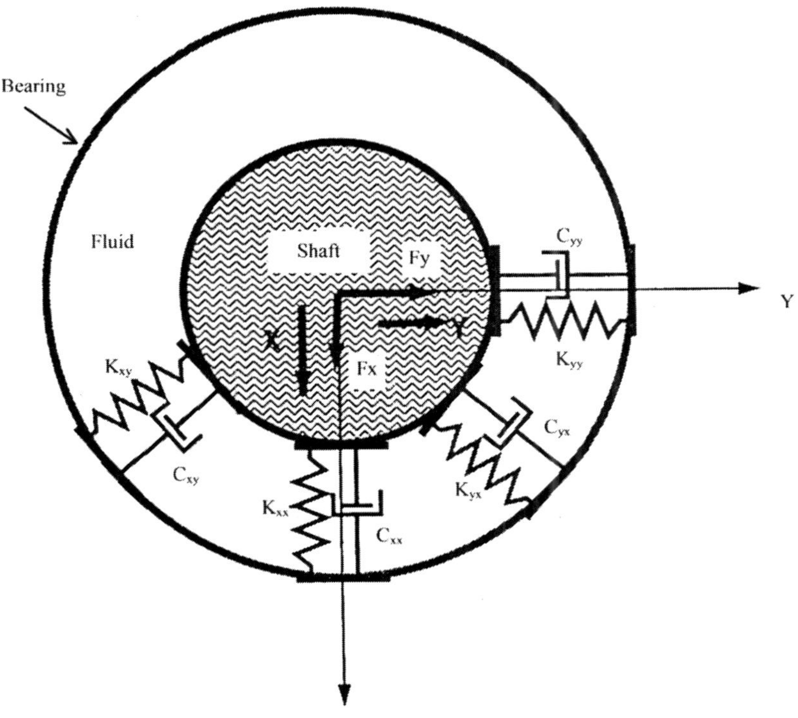

Figure 2. Linearized stiffness and damping coefficients of the journal bearing.

The eigenvalue problem associated with Equation (1) can be reduced to a standard form, following a procedure similar to that presented by [27] . A second order state vector {X}, defined in the following form, is used to rewrite the governing equation on state variables:

$$\{X\} = \left[\{\dot{U}\}^{\mathrm{T}} \middle| \{U\}^{\mathrm{T}} \right]^{\mathrm{T}} \tag{4}$$

The free vibration problem associated with Equation (1) can be rewritten as follows

$$\left[M^*\right]\{\dot{X}\} + \left[C^*\right]\{X\} = \{0\} \tag{5}$$

where

$$\left[M^*\right] = \begin{bmatrix} [M]+[N] & 0 \\ 0 & [I] \end{bmatrix} \tag{6a}$$

$$\left[C^*\right] = \begin{bmatrix} [C] & [K] \\ -[I] & 0 \end{bmatrix} \tag{6b}$$

where [I] is the identity matrix, with the same dimension as that of [M], [N], [C] and [K]. The solution of Equation (5) has the form

$$\{X\}(t) = \mathrm{e}^{st}\{X_o\}(t) \tag{7}$$

and the associated eigenvalue problem can be stated as

$$s\left[M^*\right]\{X_o\} + \left[C^*\right]\{X_o\} = \{0\} \tag{8}$$

or

$$\left[A^*\right]\{X_o\} = s\{X_o\} \quad , \tag{9}$$

where, provided [M*] is non-singular,

$$[A^*] = -[M^*]^{-1}[C^*] = \begin{bmatrix} -([M]+[N])^{-1}[C] & -([M]+[N])^{-1}[K] \\ [I] & 0 \end{bmatrix}. \tag{10}$$

As described before, the matrix [C] is defined as $[C]=[C_1]-\Omega\cdot[G]$, where [G] is the shaft gyroscopic matrix, while the matrix $[C_1]$ represents the bearing damping. The variable s in Equation (8) represents the system complex eigenvalues. These eigenvalues are composed of a real part "a" and an imaginary part "b", given by Equation (11).

$$s = a \pm ib \tag{11}$$

The imaginary part "b" corresponds to the system natural frequency and the real part "a" gives information on the system stability, as shown in the numerical results presented in the following sections.

3. NUMERICAL RESULTS

Numerical results of some rotor-bearing systems are obtained to validate the finite element procedure developed in this work and to perform the stability analysis of a damped gyroscopic system. In example 1, the predicted results of the natural frequencies obtained from the eigenvalue analysis of the damped gyroscopic system are compared to those in the literature. In example 2, a more complete system is analysed, including fluid-film bearing supports, and stability analysis based on the eigenvalues and the effective damping parameter. The finite element procedure and the eigenvalue calculation are both implemented in Matlab®.

3.1. Example 1

The validation of the finite element procedure begins with an example of a uniform shaft supported at its ends on identical damped bearings, presented by [32] , where the shear effects, rotatory inertia and gyroscopic effects are considered. Firstly, for the mesh convergence analysis, the system parameters are listed in Table 1.

The finite element mesh convergence analysis is done based on the natural frequencies of the rotor-bearing system, and compared to reference [32] results. The results are presented in Tables 2 and 3 and Figure 3, where “NF” stands for natural frequency and “n” is the number of elements. It is concluded that an 80-element mesh yields less than 2% error at the prediction of the first, second, third and fifth natural frequencies, and less than 2.4% error at the fourth natural frequency.

Another validation procedure is developed for a similar system. It consists of a stability analysis, using the data presented by [32] in his second example, where direct stiffness coefficients are 3.5024×10^6 N/m and the damping coefficients are variable, as shown in Table 4.

The complex eigenvalues lead to the damped natural frequencies of the system and these eigenvalues are compared to [32] , for the validation. The results for an 80-element mesh are presented in Figure 4 and Figure 5, with the natural frequencies in function of the bearing damping coefficient, where “EV” stands for “eigenvalue”. The solid, dotted and dashed lines are associated with the results presented by [32] , while the symbols indicate the predictions obtained by the finite element procedure developed in this work. The finite element predictions are represented by EV158, EV159, EV160, EV161 and EV162, which represent the 158th, 159th, 160th, 161st and 162nd system eigenvalues, respectively.

Table 1. System parameters adopted for the mesh convergence analysis.

Parameter	Description	Value	Unit
L	shaft length	1.270	m
d	shaft diameter	0.1016	m
E	shaft Young module	207×10^9	Pa
ρ	shaft specific mass	7833	kg/m³
k_b	bearings direct stiffness coefficient	10.51×10^6	N/m
c_b	bearings direct damping coefficient	17512	N·s/m

Table 2. Finite element mesh convergence analysis—Natural frequencies comparison (rpm).

	[32]	FEM-Prediction							
		$n = 5$	$n = 10$	$n = 20$	$n = 40$	$n = 60$	$n = 80$	$n = 100$	$n = 120$
1st NF	4260	4239	4396	4373	4342	4329	4323	4319	4316
2nd NF	6000	6296	6052	5983	5985	5994	6000	6004	6007
3rd NF	16,500	16,463	16,934	16,938	16,864	16,829	16,809	16,796	16,788
4th NF	45,000	46,557	46,580	46,444	46,202	46,091	46,031	45,992	45,966
5th NF	90,000	92,700	91,943	91,848	91,631	91,518	91,454	91,413	91,385

Table 3. Finite element mesh convergence analysis—Natural frequencies relative error.

	$n = 5$	$n = 10$	$n = 20$	$n = 40$	$n = 60$	$n = 80$	$n = 100$	$n = 120$
1st NF	−0.5%	3.2%	2.6%	1.9%	1.6%	1.5%	1.4%	1.3%
2nd NF	4.9%	0.9%	−0.3%	−0.2%	−0.1%	0.0%	0.1%	0.1%
3rd NF	−0.2%	2.6%	2.7%	2.2%	2.0%	1.9%	1.8%	1.7%
4th NF	3.5%	3.5%	3.2%	2.7%	2.4%	2.3%	2.2%	2.1%
5th NF	3.0%	2.2%	2.1%	1.8%	1.7%	1.6%	1.6%	1.5%

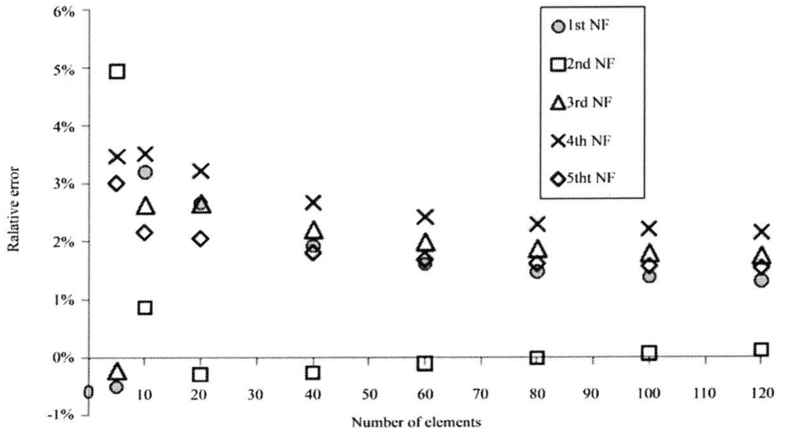

Figure 3. Figure 3. Relative error between the predicted natural frequencies and [32].

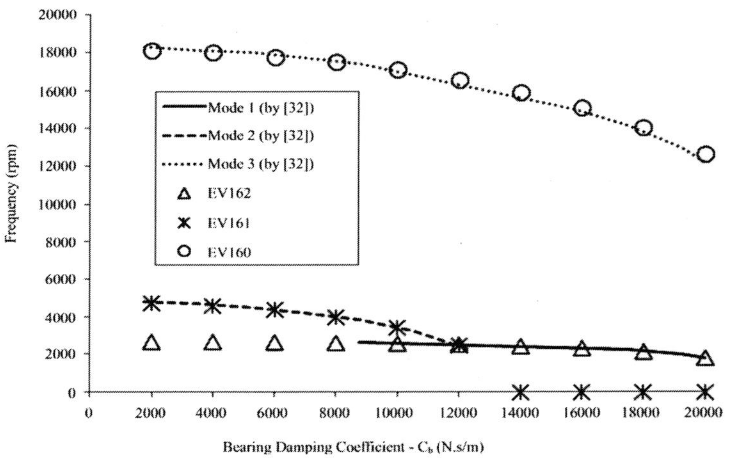

Figure 4. Natural frequencies versus bearings damping coefficients, for the 1st, 2nd and 3rd vibration modes.

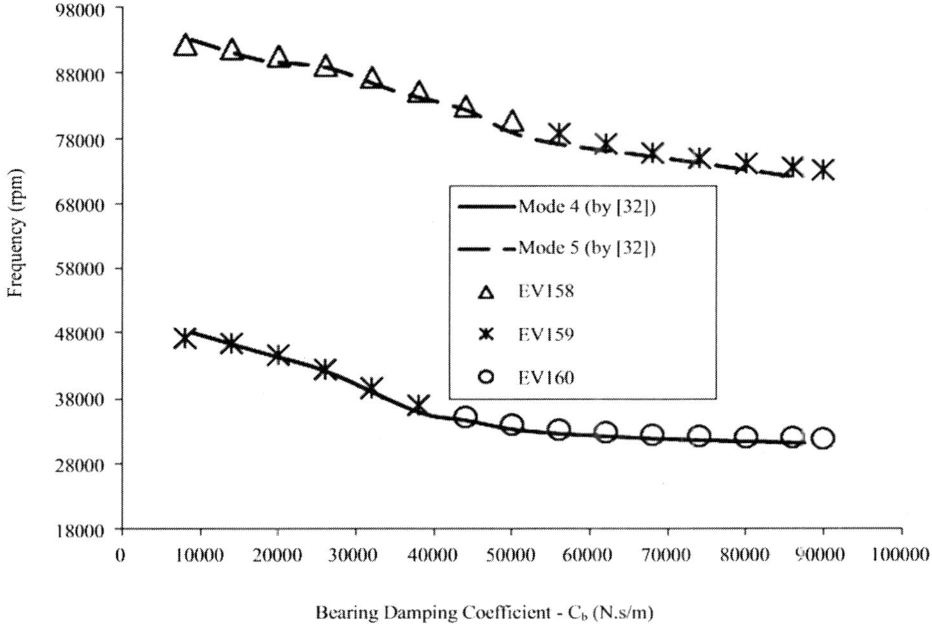

Figure 5. Natural frequencies versus bearings damping coefficients, for the 4th and 5th vibration modes.

Table 4. System parameters for stability analysis.

Parameter	Description	Value	Unit
L	Shaft length	1.270	m
d	Shaft diameter	0.1016	m
E	Shaft Young module	207×10^9	Pa
ρ	Shaft specific mass	7833	kg/m³
k_b	Bearings direct stiffness coefficient	3.5024×10^6	N/m
c_b	Bearings direct damping coefficient	-	N·s/m

The numerical procedure using state variable form produces many eigenvalues, which are separated and organized. For this example, using a mesh with 80 beam elements, after filtering and organizing the eigenvalues, it can be observed that the 160th, 161st and 162nd eigenvalues represent, respectively, the 3rd, 2nd and 1st vibration modes of the rotor-bearing system (Figure 4). As the damping coefficient increases, the order of eigenvalues corresponding to the modeshapes may change, as shown in Figure 5. The comparative results depicted on Figures 4 and 5 show that the finite element procedure renders results in good agreement with those presented by [32] , with a maximum 2% relative error.

This analysis shows the influence of the bearing damping on the natural frequencies of the rotor-bearing system. If the bearing parameters chanJge, the damping may vary, changing the vibration behaviour of the entire system. The stability can also be studied based on the eigenvalues. The computation of the damping exponent associated with part "a" of the complex eigenvalues (Equation (11)), renders results very similar to those obtained by [32] , whose results are omitted here for brevity. An analysis of the real part of the complex eigenvalues is carried out for another system and presented in example 2.

3.2. Example 2

This second example consists of a rotor-bearing system composed of a shaft supported by two hydrodynamic journal bearings, similar to that shown in Figure 1. More details can be seen in [6] . The parameters of the system are shown in Table 5.

The exponents associated with part "a" (Equation (11)) of the complex eigenvalues are computed for the rotating system with some sets of parameters associated with the hydrodynamic journal bearings. This analysis is performed to show the importance of the bearing effective damping $2\omega C_{xx}/K_{xy}$ on the stabilization of the rotating system [2] . The shaft rotating speed is represented by ω, while C_{xx} is the bearing direct damping coefficient and K_{xy} is the bearing cross-coupled stiffness coefficient.

For this example, three sets of rotating speeds have been selected (750 rpm, 3200 rpm and 5600 rpm), as shown in Table 6, in order to show how the bearing effective damping and the value of "a" are related to the system stability. The bearing stiffness coefficients remain constant, while the damping coefficients are adjusted

Table 5. System parameters for example 2.

Parameter	Description	Value	Unit
L	Shaft length	0.30	m
d	Shaft diameter	0.015	m
l	Bearings length	0.012	m
c_1	Bearing radial clearance	34.5×10^{-6}	m
μ	Lubricant viscosity	25×10^{-3}	Pa·s
ρ	Lubricant mass density	892	kg/m³
E	Shaft Young modulus	200×10^{9}	Pa
ν	Shaft Poisson coefficient	0.30	-
ρ	Shaft mass density	7870	kg/m³

Table 6. Numerical results for example 2 and stability parameters.

Speed (rpm)	$[K_m]$ (N/m)	$[C_m]$ (N·s/m)	$2\omega C_{xx}/K_{xy}$	Stability	a
750	$10^4 \times \begin{bmatrix} 4.4150 & 4.4137 \\ -4.4137 & 4.4150 \end{bmatrix}$	$\begin{bmatrix} 29.81 & 0 \\ 0 & 29.81 \end{bmatrix}$	0.1	UNSTABLE	156
750	$10^4 \times \begin{bmatrix} 4.4150 & 4.4137 \\ -4.4137 & 4.4150 \end{bmatrix}$	$\begin{bmatrix} 1579.9 & 0 \\ 0 & 1579.9 \end{bmatrix}$	5.3	Stable	−260
3200	$10^6 \times \begin{bmatrix} 0.8111 & 1.3896 \\ -1.3896 & 0.8111 \end{bmatrix}$	$\begin{bmatrix} 208.4 & 0 \\ 0 & 208.4 \end{bmatrix}$	0.1	UNSTABLE	298
3200	$10^6 \times \begin{bmatrix} 0.8111 & 1.3896 \\ -1.3896 & 0.8111 \end{bmatrix}$	$\begin{bmatrix} 4168 & 0 \\ 0 & 4168 \end{bmatrix}$	2.0	Stable	−193
5600	$10^6 \times \begin{bmatrix} 0.4092 & 2.3518 \\ -2.3518 & 0.4092 \end{bmatrix}$	$\begin{bmatrix} 200.2 & 0 \\ 0 & 200.2 \end{bmatrix}$	0.1	UNSTABLE	1767
5600	$10^6 \times \begin{bmatrix} 0.4092 & 2.3518 \\ -2.3518 & 0.4092 \end{bmatrix}$	$\begin{bmatrix} 4004 & 0 \\ 0 & 4004 \end{bmatrix}$	2.0	Stable	−130

for each rotating speed. For simplicity, the cross-coupled damping coefficients are neglected in this analysis.

The results obtained from this stability analysis show that an increase on the direct damping coefficient can stabilize a rotor operating unstably. A sign change on the eigenvalue exponent "a" indicates that the rotor is moving from unstable to stable conditions or vice-versa. The stability analysis based on the real part "a" of the complex eigenvalues associated with the rotor shown in Table 5 provides the same results as those based on the time integration of the rotor governing equation, presented by [6] . It is clear from this analysis that the bearing effective damping $2\omega C_{xx}/K_{xy}$ is a very important parameter for rotors operating on hydrodynamic bearings. When the effective damping decreases and approaches 0, the system tends to be unstable, and when it increases, the system tends to be stable. On Table 6, a system is considered unstable when $a > 0$, and stable when $a \leq 0$. The values shown on the second and third columns labelled $[K_m]$ and $[C_m]$ are the bearing dynamic force coefficients.

The journal dynamic coefficients play a crucial role on the stability of the system, as shown in this second example. By changing the damping coefficients, a stable condition can be established, as shown on Table 6.

4. CONCLUSIONS

The appropriate selection of a rotor supporting system is a fundamental step on the design and commissioning of industrial rotating machines. Dynamic force coefficients play a crucial role in the rotor capability to bear undesirable vibrations and to run under stable conditions. The results presented in this work show clearly the importance of selecting the appropriate bearing configuration able to provide enough effective damping to bound the growth of the vibration response at critical operating conditions.

The finite element procedure has been implemented to analyze the stability of high speed turbomachines supported by hydrodynamic journal bearings. The numerical results presented in this work show that the computational procedure implemented for the eigenvalue analysis of damped gyroscopic systems is able to render reliable results, which are in good agreement with the results presented in the technical literature.

The finite element procedure can also be employed to evaluate design and operating changes in high-speed turbomachinery, in order to improve their dynamic response. From the parameters of the rotor-bearing system, its dynamic behaviour can be studied and modified, for example, to avoid its operation near a critical speed, or to guarantee safe operation when traversing critical speeds.

REFERENCES

1. Childs, D.W. (1993) Turbomachinery Rotordynamics. Wiley, New York, 476 p.
2. Vance, J. (1988) Rotordynamics of Turbomachinery. John Wiley, New York, 388 p.
3. Nelson, H.D. and McVaugh, J.M. (1976) The Dynamics of Rotor-Bearing Systems Using Finite Elements. ASME Journal of Engineering for Industry, 98, 593-600.
4. Zorzi, E.S. and Nelson, H.D. (1977) Finite Element Simulation of Rotor-Bearing Systems with Internal Damping. ASME Journal of Engineering for Power, 99, 71-76.

5. Nelson, H.D. (1980) A Finite Rotating Shaft Element Using Timoshenko Beam Theory. ASME Journal of Mechanical Design, 102, 793-803.

6. Miranda, W.M., Machado, L.H.J. and Faria, M.T.C. (2005) Some Insights into the Dynamic Response of Flexible Rotors Supported on Fluid Film Bearings. 18th International Congress of Mechanical Engineering, COBEM 2005, Ouro Preto, 8 p.

7. Correia, F.A.G. (2007) Determinação de Características de Desempenho de Mancais Radiais Elípticos Utilizando o Método de Elementos Finitos. Masters' Dissertation, Universidade Federal de Minas Gerais, Brasil.

8. Busse, L., Heiberger, D. and Wey, J. (1980) Aspects of Shaft Dynamics for Industrial Turbines. Brown Boveri Review, 67, 292-299.

9. Sternlicht, B. and Lewis, P. (1968) Vibration Problems with High-Speed Turbomachinery. ASME Journal of Engineering for Industry, 90, 174-186.

10. Lund, J.W. (1994) Application of Modal Methods in Rotor Dynamics. 4th IFTOMM Conference—Rotordynamics, Chicago, 317-320.

11. Boedo, S. and Booker, J.F. (1997) Surface Roughness and Structural Inertia in a Mode-Based Mass-Conserving Elastohydrodynamic Lubrication Model. ASME Journal of Tribology, 119, 449-455.

12. Zhao, S., Xu, H., Meng, G. and Zhu, J. (2005) Stability and Response Analysis of Symmetrical Single-Disk Flexible Rotor-Bearing System. Tribology International, 38, 749-756.

13. Xie, W.H., Tang, Y.G. and Chen Y.S. (2008) Analysis of Motion Stability of the Flexible Rotor-Bearing System with Two Unbalanced Disks. Journal of Sound and Vibration, 310, 381-393.

14. Gupta, K.K. (1974) Eigenproblem Solution of Damped Structural Systems. International Journal for Numerical Methods in Engineering, 8, 877-911.

15. Meirovitch, L. and Ryland, G. (1979) Response of Slightly Damped Gyroscopic Systems. Journal of Sound and Vibration, 67, 1-19.

16. Palazzolo, A.B., Wang, B.P. and Pilkey, W.D. (1983) Eigensolution Reanalysis of Rotor Dynamic Systems by the Generalized Receptance Method. ASME Journal of Engineering for Power, 105, 543-550.

17. Ehrich, F.F. (1992) Handbook of Rotordynamics. McGraw-Hill, New York, 452 p.

18. Li, L., Hu, Y. and Wang, X. (2012) A Parallel Way for Computing Eigenvector Sensitivity of Asymmetric Damped Systems with Distinct and Repeated Eigenvalues. Mechanical Systems and Signal Processing, 30, 61-77.

19. Plaut, R.H. and Huseyin, K. (1973) Derivatives of Eigenvalues and Eigenvectors in Non-Self-Adjoint Systems. AIAA Journal, 11, 250-253.

20. Lund, J.W. (1980) Sensitivity of the Critical Speeds of a Rotor to Changes in the Design. ASME Journal of Mechanical Design, 102, 115-121.

21. Rajan, N., Nelson, H.D. and Chen, W.J. (1986) Parameter Sensitivity in the Dynamics of Rotor-Bearing Systems. ASME

Journal of Vibration, Acoustics, Stress, and Reliability in Design, 108, 197-206.

22. Done, G.T.S. and Hughes, A.D. (1975) The Response of a Vibrating Structure as a Function of Structural Parameters. Journal of Sound and Vibration, 38, 255-266.

23. Rao, B.S., Sekhar, A.S. and Majumdar, B.C. (1996) Analysis of Rotors Considering Distributed Bearing Stiffness and Damping. Compurers & Structures, 61, 951-955.

24. Werner, U. (2013) Mathematical Rotordynamic Model Regarding Excitation Due to Elliptical Shaft Journals in Electrical Motors Considering the Gyroscopic Effect. Applied Mathematics, 4, 57-74.

25. Jeon, S.M., Kwak, H.D., Yoon, S.H. and Kim, J. (2013) Rotordynamic Analysis of a High Thrust Liquid Rocket Engine Fuel (Kerosene) Turbopump. Aerospace Science and Technology, 26, 169-175.

26. Meirovitch, L. (1974) A New Method of Solution of the Eigenvalue Problem for Gyroscopic Systems. AIAA Journal, 12, 1337-1342.

27. Meirovitch, L. (1980) Computational Methods in Structural Dynamics. Sijthoff & Noofdhoff International Publishers, The Netherlands.

28. Childs, D.W. and Graviss, K. (1982) A Note on Critical Speed Solutions for Finite-Element-Based Rotor Models. ASME Journal of Mechanical Design, 104, 412-416.

29. Faria, M.T.C. (2001) Some Performance Characteristics of High Speed Gas Lubricated Herringbone Groove Journal Bearings. JSME International Journal, Series C, 44, 775-781.

30. Faria, M.T.C., Correia, F.A.G. and Teixeira, N.S. (2007) Oil Lubricated Cylindrical Journal Bearing Analysis Using the Finite Element Method. Proceeding of the 19th International Congress of Mechanical Engineering (COBEM 2007), Brasilia, 10 p.

31. James, M.L., Smith, G.M., Wolford, J.C. and Whaley, P.W. (1994) Vibration of Mechanical and Structural Systems. Harper Collins, 2nd Edition, USA, 660 p.

32. Lund, J.W. (1974) Stability and Damped Critical Speeds of a Flexible Rotor in Fluid Film Bearings. ASME Journal of Engineering for Industry, 96, 509-517.

9

CHAPTER

ANALYTICAL ESTIMATION OF ELASTIC PROPERTIES OF POLYPROPYLENE FIBER MATRIX COMPOSITE BY FINITE ELEMENT ANALYSIS

Bhaskar Pal[1], Mohamed Riyazuddin Haseebuddin[2]

[1]Mechanical Engineering Department, Dayananda Sagar Academy of Technology and Management, Bangalore, India

[2]Mechanical Engineering Department, Dayananda Sagar College of Engineering, Bangalore, India

ABSTRACT

A structural composite is a material system consisting of two or more phases on a macroscopic scale, whose mechanical performance and properties are designed to be superior to those of constituent materials acting independently. Fiber reinforced composites (FRP) are slowly emerging from the realm of advanced materials and are replacing conventional materials in a variety of applications. However, the mechanics of FRPs are complex owing to their anisotropic and heterogeneous characteristics. In this paper a representative volume model has been considered and a finite element model incorporating the necessary boundary conditions is developed using available FEA package ANSYS to predict the elastic property of the composite. For verification, the numerical results of elastic properties are compared with the analytical solution and it is found that there is a good agreement between these results.

Keywords: FRP; RVE; Finite Element Method; Homogenization

1. INTRODUCTION

In recent years, there has been a rapid growth in the use of fiber-reinforced composites due to their ability to replace competitive materials on the basis of lower density and equivalent strength, low thermal conductivity, high corrosion and wear resistance and the possibility of combining the toughness of thermoplastic polymers with the stiffness and strength of reinforcing fibres. This has resulted in the need for variety of application like general, structural, automotive and aerospace industry.

Unidirectional composites are those which have all fibres aligned in a single direction. A unidirectional composite with a hexagonal array of fibres can be transversely isotropic because the properties are the same along any plane which is normal to the fibre direction [1]. The stiffness and strength of a unidirectional composite are anisotropic properties since they vary with orientations. The stiffness of unidirectional composites in the fibre direction is usually dominated by the fibre properties while the strength in the transverse direction is dominated by the matrix properties. Since the strength of a unidirectional composite under transverse tension is much smaller than under longitudinal tension, transverse tensile loading is believed to be the critical loading of unidirectional composite materials [2].

Composite elastic properties are determined by the physical and mechanical properties of the individual materials. Some analytical and numerical techniques have been used for prediction and characterization of composite behaviour. Analytical methods provide reasonable prediction for relatively simple configurations of the phases. Complicated geometries, loading conditions and material properties often do not yield analytical solutions, due to complexity and the number of equations. In this case, numerical methods are used for approximate solutions, but they still make some simplifying assumptions about the inherent microstructures of heterogeneous multiphase materials, one such method is finite element analysis [3-5].

One of the basic requirements in the mechanics of composite materials is to determine the effective elastic properties. Experimental determination of the moduli of unidirectional composites is difficult, especially when it involves determining the longitudinal shear and transverse shear moduli. Thus, numerical techniques like the finite element method are needed to calculate these shear moduli. Numerical methods to calculate composite material properties usually involve analysis of a representative volume element (RVE) [6-8].

A great number of micromechanical models have been proposed in the literature for predicting various mechanical properties of composite materials [9-11]. Several other models have been proposed such as numerical homogenization [9], FEM.

Among all the theoretical models, the ROM has got the simplest mathematical relations. To apply these models, the modulus of elasticity of the polymer, E_m, and of the fiber, E_f, should be known and then the modulus of elasticity of the composite, $E_{1,2}$, can be calculated for any volume fraction of the fiber in the composition.

But those experienced in the field shall admit that this model, in most cases, do not predict the modulus of elasticity of the composites satisfactorily. The experimental observations and analysis also confirm that [5,12-14].

There are a few issues that need to be verified carefully when carrying out such analyses. Firstly, the correct RVE corresponding to the assumed fibre distribution must be isolated. Secondly, correct boundary conditions need to be applied to the chosen RVE to model different loading situations. Proper consideration must be given to the periodicity and symmetry of the model in arriving at the correct boundary condition.

In the present work, the procedure for predicting the elastic constants of the composite from the RVE is established for a micromechanical three-dimensional finite element analysis. The finite element calculations are made at a specific volume fraction because the geometry of the regions of the finite-element mesh that represent the fibre and matrix differ from one volume fraction to the next and, therefore, each volume fraction study requires a separate analysis. The finite element method is adopted for predicting various elastic properties of unidirectional oriented FRP and the results of E_1, E_2, v_{12} and v_{23} are compared with the rule of mixtures and HalphinTsai criteria.

2. METHODOLOGY

2.1. Analytical Formulation

2.1.1. Role of Mixture

Rules of Mixtures are mathematical expressions which give some property of the composite in terms of the properties, quantity and arrangement of its constituents. The notations will be established by use of the following relationships:

- Longitudinal Young's Modulus

$$E_1 = E_f V_f + E_m (1 - V_f) \tag{1}$$

- Transverse Young's Modulus

$$\frac{1}{E_2} = \frac{V_f}{E_f} + \frac{(1 - V_f)}{E_m} \tag{2}$$

- Major Poisson's Ratio

$$v_{12} = v_f V_f + v_m (1 - V_f) \tag{3}$$

- In-Plane Shear Modulus

$$\frac{1}{G_{12}} = \frac{V_f}{G_f} + \frac{(1 - V_f)}{G_m} \tag{4}$$

2.1.2. Semi-Empirical Model (Halphin-Tsai)

Halphin and Tsai developed their models as simple equations by curve fitting to results that are based on elasticity. The equations are semi-empirical in nature since involved parameters in the curve fitting carry physical meaning.

- Longitudinal Young's Modulus

$$E_1 = E_f V_f + E_m V_m \tag{5}$$

- Transverse Young's Modulus

$$\frac{E_2}{E_m} = \left[\frac{1+\xi\eta V_f}{1-\eta V_f}\right], \quad \eta = \left[\frac{\left[\frac{E_f}{E_m}\right]-1}{\left[\frac{E_f}{E_m}\right]+\xi}\right] \tag{6}$$

The term "ξ" is called the reinforcing factor and depends on the following;

Fiber geometry, Packing geometry, Loading condition

- Major Poisson's Ratio

$$v_{12} = v_f V_f + v_m V_m \tag{7}$$

- In-Plane Shear Modulus

$$\frac{G_{12}}{G_m} = \left[\frac{1+\xi\eta V_f}{1-\eta V_f}\right], \quad \eta = \left[\frac{\left[\frac{G_f}{G_m}\right]-1}{\left[\frac{G_f}{G_m}\right]+\xi}\right] \tag{8}$$

2.2. Compliance Matrix

In composite material fibers may be oriented in an arbitrary manner. Depending on the arrangements of the fibers, the material may behave differently in different directions. According to their behaviour, composites may be characterized as generally anisotropic, monoclinic, orthotropic, and transversely isotropic. In this paper, transversely isotropic characteristics have been considered for the fiber reinforced composite and fiber arrangement as shown in the Figure 1.

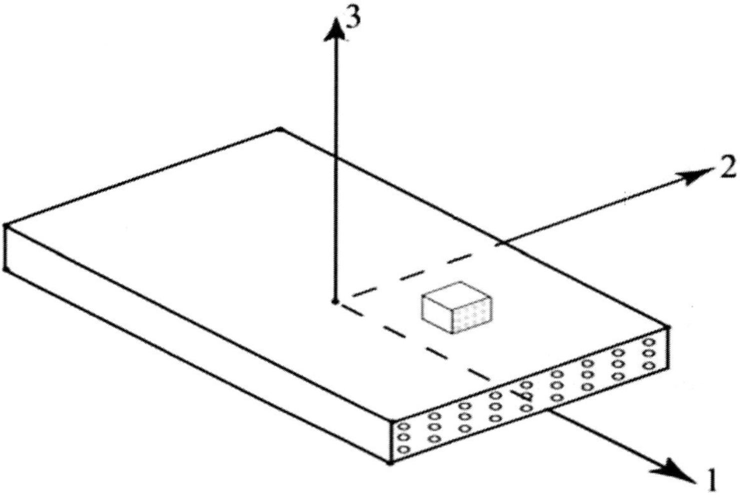

Figure 1. Arrangement of fiber direction for transversely isotropic composite.

$$\begin{bmatrix} \epsilon_x \\ \epsilon_y \\ \epsilon_z \\ \gamma_{23} \\ \gamma_{13} \\ \gamma_{12} \end{bmatrix} = \begin{bmatrix} 1/E_1 & -\gamma_{12}/E_2 & -\gamma_{12}/E_2 & 0 & 0 & 0 \\ -\gamma_{12}/E_1 & 1/E_2 & -\gamma_{23}/E_2 & 0 & 0 & 0 \\ -\gamma_{12}/E_1 & -\gamma_{23}/E_2 & 1/E_2 & 0 & 0 & 0 \\ 0 & 0 & 0 & 2(1+\gamma_{23})/E_2 & 0 & 0 \\ 0 & 0 & 0 & 0 & 1/G_{12} & 0 \\ 0 & 0 & 0 & 0 & 0 & 1/G_{12} \end{bmatrix} \times \begin{bmatrix} \sigma_x \\ \sigma_y \\ \sigma_z \\ \tau_{23} \\ \tau_{13} \\ \tau_{12} \end{bmatrix}$$

Here 1-2-3 orthogonal coordinate system is used where the directions are taken as follows:

- The 1-axis is aligned with the fiber direction.
- The 2-axis is in the plane of the layer and perpendicular to the fibers.
- The 3-axis is perpendicular to the plane of the layer and thus also perpendicular to the fibers.

Stress strain relationship for compliance matrix for transversely isotropic matrix given below.

2.3. Representative Volume Element (RVE)

Primary to use of numerical approximations of the effective properties of composite is the concept of representative volume element (RVE). Square or cubic RVEs are used for most numerical approximations because of the ease of numerically solving boundary values problems with these geometries. The difficulties involved in generating statistical information about particle distributions and concentrations leads to difficulties in the rigorous determination of RVE sizes. Hence, for most applications, RVE sizes have been rather arbitrary. In this paper a RVE model of 420 μm × 420 μm has considered which consists of different volume fraction of fiber density. In the model calculations, cuboid will be matrix and cylinder will be fiber. Figure 2 shows the typical RVE which consists of 10%, 17%, 27%, 40% and 50% of fiber content.

2.4. Homoginization

In classical lamination theory the composite lamina is modelled as a homogeneous orthotropic medium with certain effective moduli that describe the "average" material properties of the composite. To describe this macroscopically homogeneous medium, macro-stress and macro-strain are derived by averaging the stress and strain tensor over the volume of the RVE. The average stress and strain quantities defined in Equations (9) and (10) thus ensure equivalence in strain energy between the equivalent homogeneous material and the original heterogeneous material. These average quantities will be used in the subsequent analysis to determine composite moduli

$$\overline{\sigma_{ij}} = \frac{1}{V}\int_{v} \sigma_{ij} \cdot dV \tag{9}$$

$$\overline{\in_{ij}} = \frac{1}{V}\int_{v} \in_{ij} \cdot dV \tag{10}$$

3. COMPUTATIONAL DETAILS

In this present work, finite element method is used to approximate the different elastic property of the fiber reinforced composites by ANSYS 12.

Assumptions made for the present analysis were

- The composite is
  - ◆ Macroscopically homogeneous;
  - ◆ Linearly elastic;
  - ◆ Macroscopically transversely isotropic;
  - ◆ Initially stress free (no thermal stress).
- The fibers are:
  - ◆ Homogeneous;
  - ◆ Linearly elastic;
  - ◆ Isotropic;
  - ◆ Regularly spaced;
  - ◆ Perfectly aligned.
- The matrix is:
  - ◆ Homogeneous;
  - ◆ Linearly elastic;
  - ◆ Isotropic.

3.1. Modelling

A regular three-dimensional arrangement of short fibre in a matrix was adequate to describe the overall behaviour of the composite, was modelled as a regular uniform arrangement, as shown in Figure 1. This model assumed that the fibre was a perfect cylinder of length 250 μm, and diameter (d = 50 μm) in a cube (420 × 420 × 250 μm^3) of matrix. It is assumed that the geometry, material and loading of the unit cell are symmetrical with respect to x-y-z coordinate system as shown in Figure 2. Therefore, 10% volume fraction i.e. 9 fibers has been inserted in a cubic matrix (420 × 420 × 250 μm^3) uniformly as shown in Figure 2. 3D finite element meshing for the 10% volume fraction of the fiber has been done, similarly the finite element modelling and meshing were done by varying the fiber volume fraction from 10% to 54% and as shown in Figure 3.

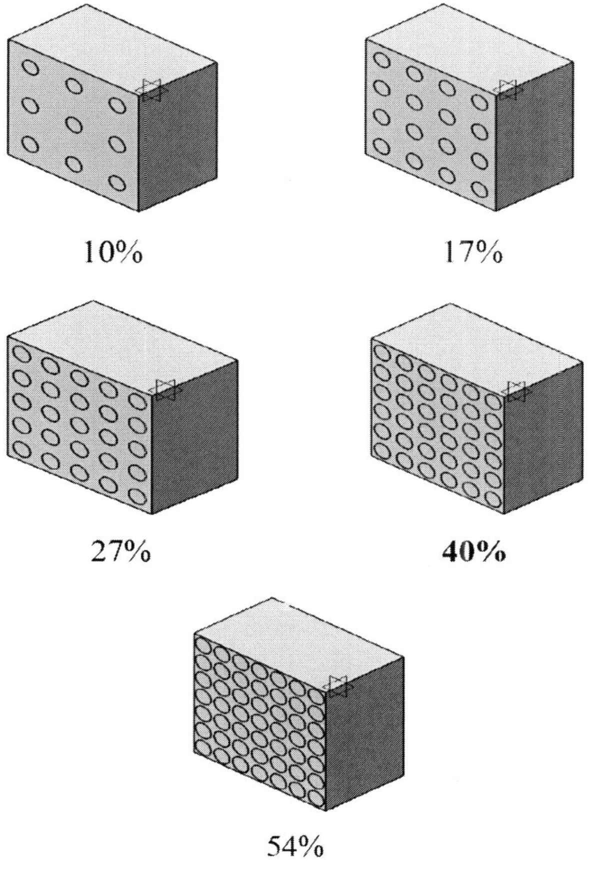

Figure 2. Unit cell of square array fibre packing geometry for RVE model.

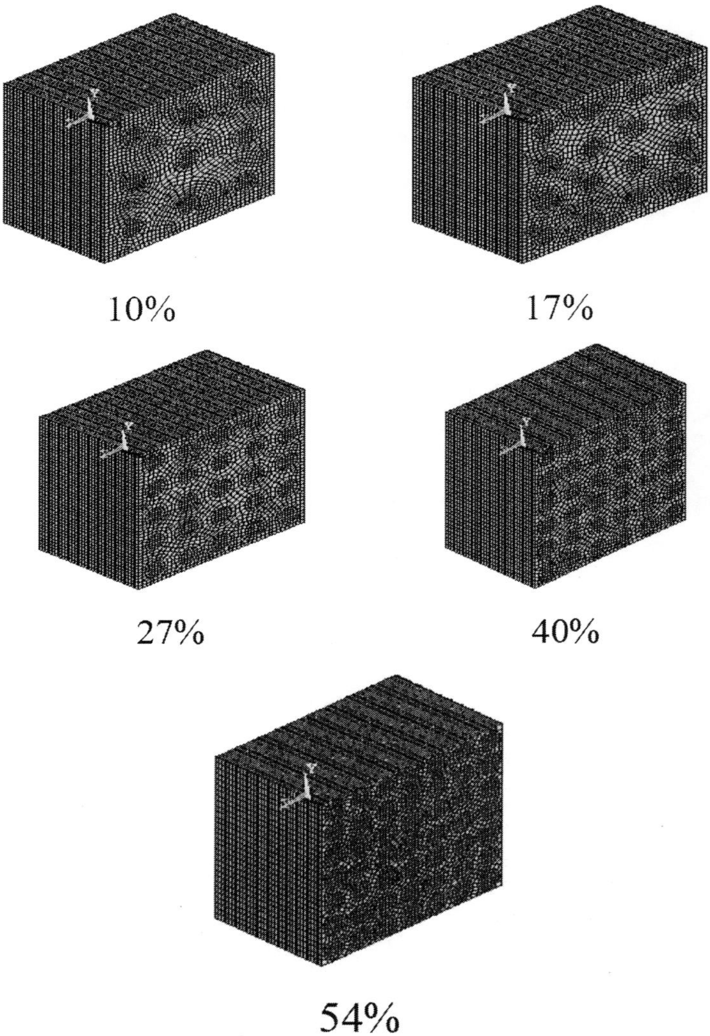

Figure 3. Finite element mesh for 10% volume fraction.

3.2. Element Type

The element SOLID 45 of ANSYS V12.0 used for the present analysis which is based on a general 3D state of stress and is suited for modeling 3D solid structure under 3D loading. The element has 8 noded brick element with three degrees of freedom per node (UX, UY and UZ).

3.3. Boundary Condition

In this work the boundary condition with normal strain applied in x direction are as follows [11]:

$$u(LF) = 0,\ v(BF) = 0,\ w(BKF) = 0 \text{ and } u(RF) = \delta$$

where LF, BF, BKF and RF stand for left face, bottom face, back face and right face of the RVE model. All other faces are free of any displacement.

In this paper, axial loading is modelled by a displacement acting on the plane yz at x. For such loading conditions, the boundaries of the RVE also correspond to lines of symmetry. Thus, normal displacements of the boundaries of the quadrant are restricted to those that cause the boundary to displace only parallel to the original boundary.

The displacement constraints applied to the finite element model to determine E_1 are as follows and as shown in Figure 4.

$$u_x(0,y,z)=0$$
$$u_x(x,y,z)=\delta_x=5.0E-05$$
$$u_y(x,0,z)=0$$
$$u_y(x,y,z)=\text{constant}=\delta_y$$
$$u_z(x,y,0)=0$$
$$u_z(x,y,z)=\text{constant}=\delta_z$$

3.4. Material Property

The different material properties of both matrix and fiber has shown in the Table 1.

Material properties are used to determine the several elastic properties of the composite that is Longitudinal young's modulus (E_1) and Transverse young's modulus (E_2) by varying the volume fraction of the fiber.

3.5. Sample Analysis

Figure 5 shows the analysis of 40% of volume fraction of fiber reinforced composite in which the von misses stresses for the matrix fiber and composite has been indicated. The stresses are obtained by applying the displacement loading in longitudinal direction.

4. RESULTS AND DISCUSSION

4.1. Longitudinal Young's Modulus (E_1)

Figure 6 shows comparison of finite element data, rule of mixtures and Halphin-Tasi results for composite modulli E_1 at different volume fractions. The finite-element solution gives identical results to the rule of mixtures and semi empirical analytical formulation. The linear dependence of E1 on fibre volume fraction is demonstrated and, as expected, the modulus increases while increasing the fibre volume fraction.

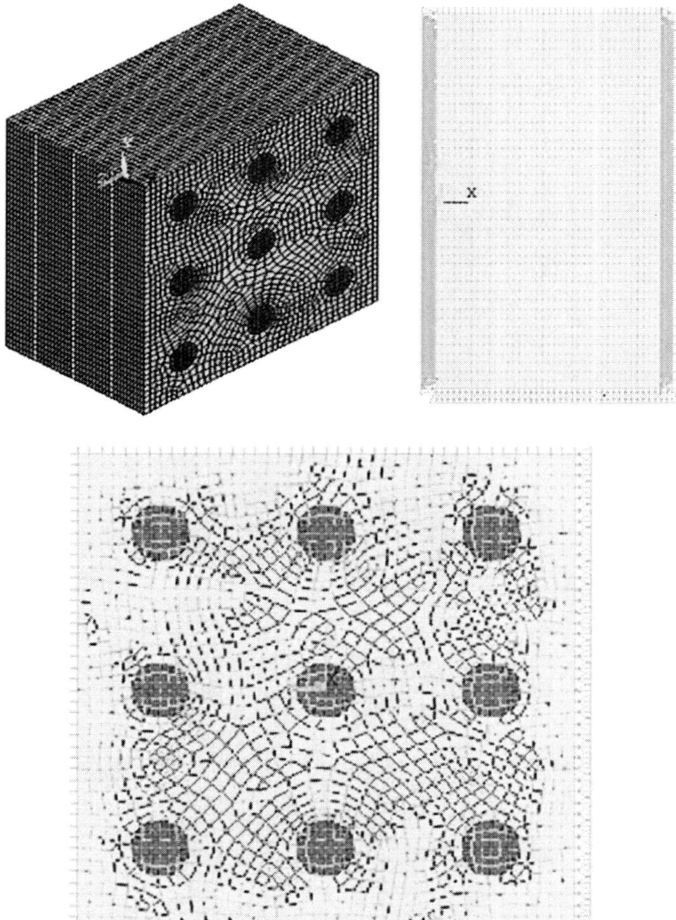

Figure 4. Boundary condition for longitudinal modulus of composite in x direction.

Table 1. Shows the material properties of composites.

Material Properties	Polypropylene co ethylene (PPE)	Polypropylene (PP)
Longitudinal Modulus E_1	1.05 Gpa	4.5 Gpa
Transverse Modulus E_2	1.05 Gpa	4.5 Gpa
Poisson's Ratio	0.33	0.2

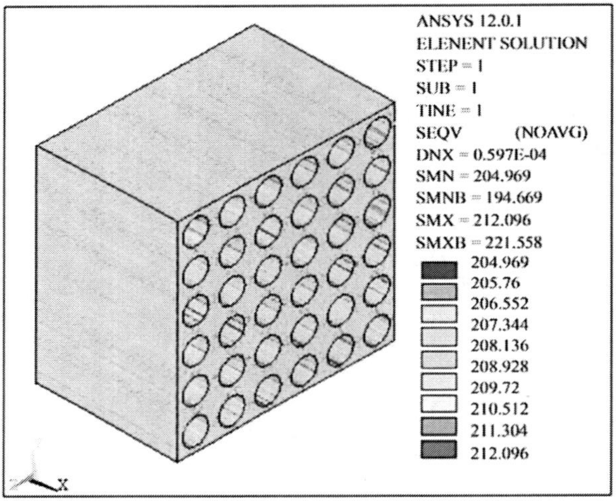

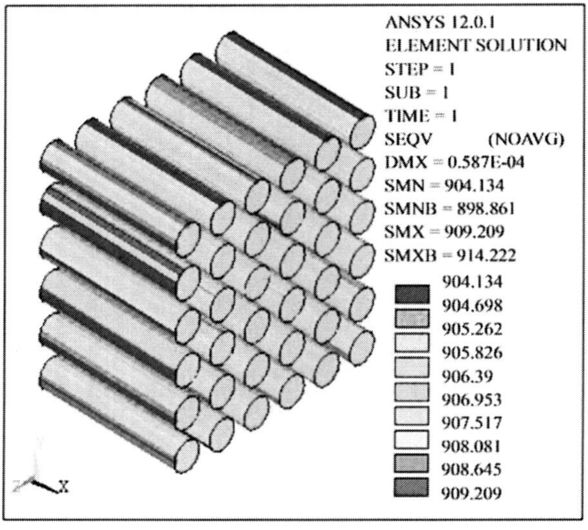

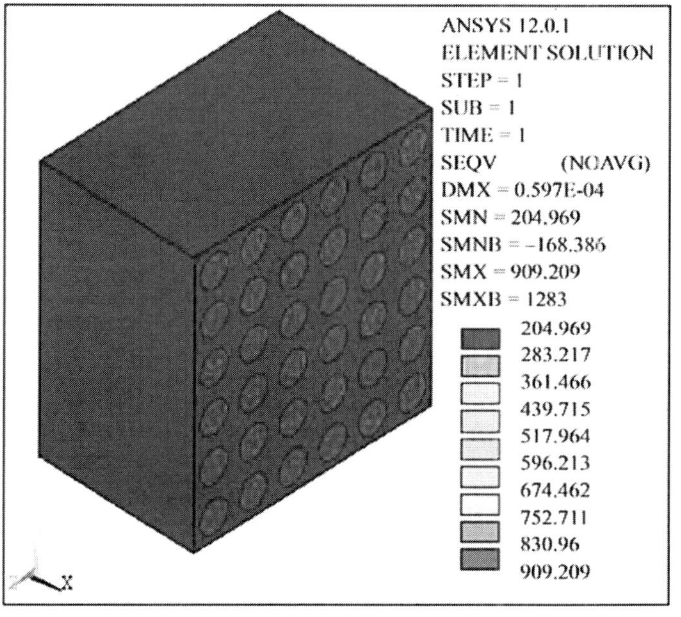

Figure 5. Stress distribution of matrix, fiber and composite during computing of E_1.

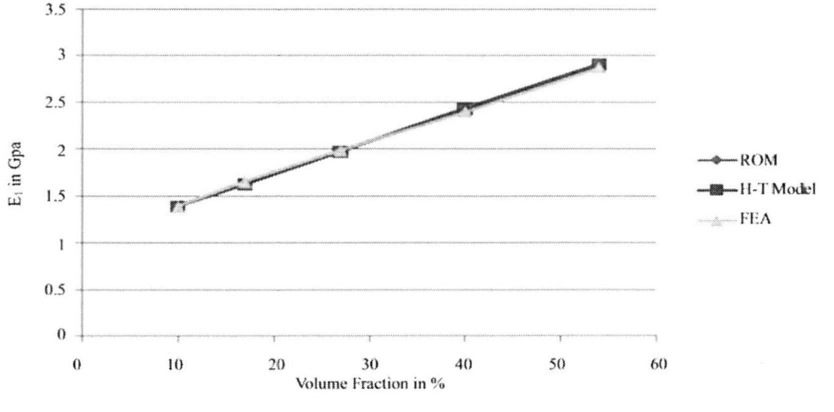

Figure 6. Comparison of finite element data, rule of mixtures and Halphin-Tasi results for composite modulli E_1 for different volume fration of fiber.

4.2. Transverse Young's Modulus (E_2)

Figure 7 shows comparison of finite element data, rule of mixtures and Halphin-Tasi results for composite modulli E_2 at different volume fractions. Finite element results are as expected; on the assumption that the composite is macroscopically transversely isotropic the values obtained for the whole unit cells investigated are perfectly coincident. The Rule of mixtures model, the Halpin-Tsai equation and data from the finite-element calculations are compared. The term "ξ" is called the reinforcing factor in Halpin Tsai equation and depends on the fiber geometry, packing geometry, loading condition. As the fibers are circular and packing density is increasing accordingly the "ξ = 2" is adapted to consider those changes in the composites. In this Halpin-Tsai equation shows a good agreement between finite element values. But in Rule of mixture there is no such parameter which can be considered for those changes. Due to this reason The Rule of mixture is not showing the good agreement with finite element data.

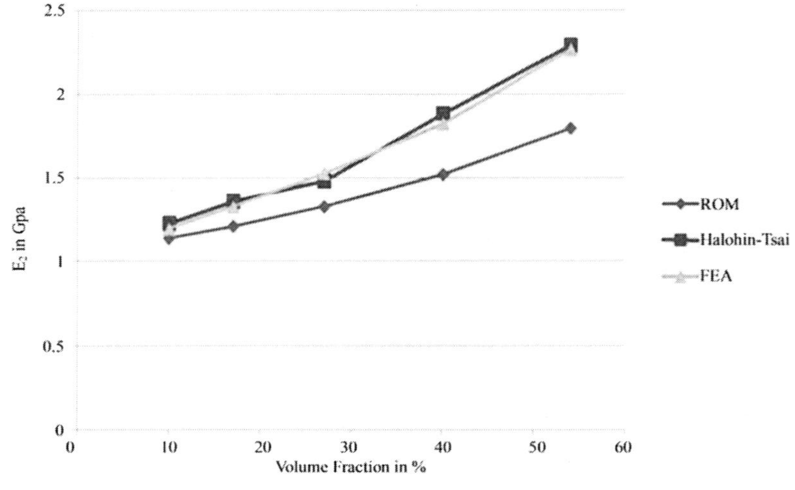

Figure 7. Comparison of finite element data, rule of mixtures and Halphin-Tasi results for composite modulli E_2 for different volume fration of fiber.

4.3. Poisson's Ratio

The major Poisson's ratio ϑ_{12} for the composite is defined as minus the ratio of strain in the y-direction divided by the strain in the x-direction when only the stress σ_x is applied. The Rule of Mixtures expression for the composite Poisson's ratio is similar in form to the expression for E_1. It evident from the Figure 8 that analytical and numerical results are in perfect agreement although data from the Rules of Mixtures are slightly higher than the finite-element results.

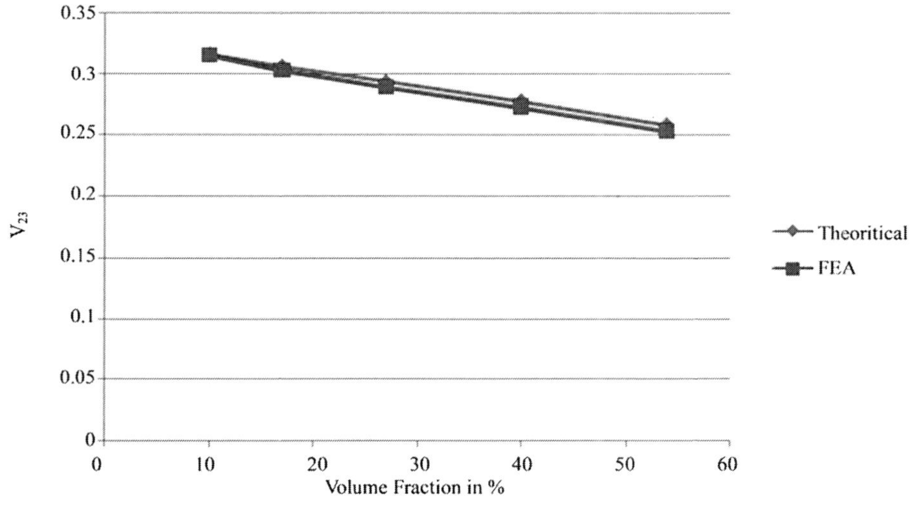

Figure 8. Comparison of finite element data and rule of mixtures results for composite poisson's ratio ϑ_{12}.

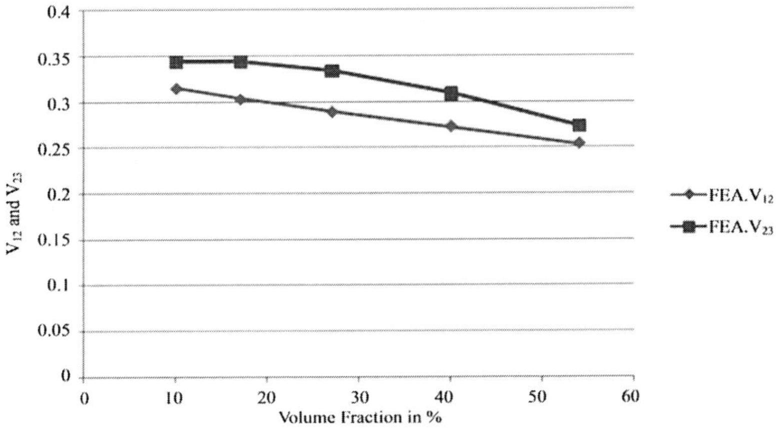

Figure 9. Comparison of finite element results for composite poisson's ratio ϑ_{12} and ϑ_{23}.

The Poisson's ratio ϑ_{23} describes the contraction in the z-direction on ϑ_{23} applying loads in the y-direction. No analytical formulas have been applied to compare with numerical results of Poisson's ratio. It is worthwhile to show finite element results of both with different fibre volume fractions. As expected from the Figure 9, ϑ_{23} appears to be higher than ϑ_{12} and the comparison between the two Poisson's ratios shows that Poisson's ratios ϑ_{12} are less sensitive to the fibre volume fraction. In particular, the values of ϑ_{12} are quite close to the Poisson's ratio of polypropylene (PP) especially at higher fibre volume fraction. On the contrary, ϑ_{23} is strongly dependent from the fibre volume fraction and, its values approaches the Poisson's ratio of the Polypropylene co ethylene at lower values of the fibre volume fraction.

5. CONCLUSIONS

1) The 3D unit cell of the unidirectional fiber reinforced composite has been identified (RVE) as a basic building block for estimating overall mechanical composite properties.

2) Various micromechanical methods to determine the elastic behaviour of composite materials have been discussed.

3) Finite element analysis has provided an implicit means of modelling polymer composites.

4) Numerical homogenization tools have been developed for the evaluation of the effective material properties of the short fiber composites.

5) The appropriate constraints on the RVE under various loadings have been determined from symmetry boundary conditions, obtaining a nearly complete set of elastic constants for a three-dimensional unidirectional composite.

6) The results of elastic moduli E_1, E_2, E_2, ν_{12}, ν_{23}, compared with the results of analytical solution and it is found that the results from FE simulation are in good agreement with the analytical results employed in this exercise namely the Halpin-Tsai semi-empirical expression and to some extent the Rules of Mixtures.

7) As this mechanical property of fiber-filled composites are affected by a number of parameters such as fiber type, matrix type, fiber orientation, fiber geometry, volume fraction of the fibers and the degree of interfacial adhesion between the fiber and the polymer matrix, the ROM has not been adopted by this property expect the volume fraction. So the results are also not so accurate.

8) The Halpin-Tsai (HT) model is also a theoretical model. This model, besides the modulus of elasticity of the polymer, E_m, and

of the fiber, E_f, includes a geometrical parameter (aspect ratio) of the fiber as well. The model has a complicated mathematical structure with fiber geometry, packing geometry, loading condition. This factors help to showing the good agreement with finite element data.

9) It is observed that the change in volume fraction of fiber has a significant effect on elastic properties.

REFERENCES

1. D. F. Adams and D. R. Doner, "Longitudinal Shear Loading of a Unidirectional Composite," Journal Composite Materials, Vol. 1, No. 1, 1967, pp. 4-17.

2. D. F. Adams and D. R. Doner, "Transverse Normal Loading of a Unidirectional Composite," Journal Composite Materials, Vol. 1, No. 2, 1967, pp. 152-164.

3. S. Houshyar, R. A. Shanks and A. Hodzic, "Modelling of Polypropylene Fibre-Matrix Composites Using Finite Element Analysis," Express Polymer Letters, Vol. 13, No. 1, 2009, pp. 2-12.

4. S. Houshyar, R. A. Shanks and A. Hodzic, "The Effect of Fibre Concentration on Mechanical and Thermal Properties of Fibre Reinforced Polypropylene Composites," Journal of Applied Polymer Science, Vol. 96, No. 6, 2005, pp. 2260-2272.

5. G. Kalaprasad, K. Joseph and S. Thomas, "Theoretical Modelling of Tensile Properties of Short Sisal Fibre-Reinforced Low Density Polyethylene Composites," Journal of Materials Science, Vol. 32, No. 16, 1999, pp. 4261- 4267.

6. C. T. Sun and R. S. Vaidya, "Prediction of Composite Properties from a Representative Volume Element," Composites Science and Technology, Vol. 56, No. 2, 1996, pp. 171-179.

7. S. Li, "General Unit Cell for Micromechanical Analyses of Unidirectional Composites," Composites Part A, Vol. 32, No. 6, 2000, pp. 815-816.

8. B. R. Kim and H. K. Lee, "An RVE-Based Micromechanical Analysis of Fiber-Reinforced Composites Considering Fiber Size Dependency," Composite Structures, Vol. 90, No. 4, 2009, pp. 418-427.

9. P. De Buhan and A. Taliercio, "A Homogenisation Approach to the Yield Strength of Compositematerials," European Journal of Mechanics, Vol. 10, No. 2, 1991, pp. 129-154.

10. Y. Pan, L. Iorga and A. A. Pelegri, "Numerical Generation of a Random Chopped Fiber Composite RVE and Its Elastic Pproperties," Composites Science and Technology, Vol. 68, No. 13, 2008, pp. 2792-2798.

11. M. Porfiri and N. Gupta, "Effect of Volume Fraction and Wall Thickness on the Elastic Properties of Hollow Particle Filled Composites," Composites Part B: Engineering, Vol. 40, No. 2, 2009, pp. 166-173.

12. K. Hbaieb, Q. Wang, Y. H. J. Chia and B. Cotterell, "Modelling Stiffness of Polymer/Clay Nanocomposites," Polymer, Vol. 48, No. 3, 2007, pp. 901-909.

13. A. G. Facca, M. T. Kortschot and N. Yan, "Predicting the Elastic Modulus of Natural Fibre Reinforced Thermoplastics," Composites Part A: Applied Science and Manufacturing, Vol. 37, No. 10, 2006, pp. 1660-1671.

14. L. Peponi, J. Biagiotti, J. M. Kenny and I. Mondragon, "Statistical Analysis of the Mechanical Properties of Natural Fibers and Their Composite Materials. II. Composite Materials," Polymer Composites, Vol. 29, No. 3, 2008, pp. 321-325.

10

CHAPTER

APPLICATION OF FINITE ELEMENTS METHOD FOR STRUCTURAL ANALYSIS IN A COFFEE HARVESTER

Evandro Pereira da Silva, Fábio Moreira da Silva, Ricardo Rodrigues Magalhães

Department of Engineering, University of Lavras, Lavras, Brazil

ABSTRACT

Stress concentration and large displacements are usual problems in the components of the structure of agricultural machinery such harvesters coffee, and that finite element method (FEM) can be a tool to minimize its effects. The goal of this paper is to get results of stresses and displacements of a coffee harvester structure by using FEM for static simulation. The main parts of the coffee harvester analyzed were: engine frame, body right and left sides, front and rear end, main beam, coffee reservoir, wheels and fuel tank. Two different design concepts of a coffee harvester machine were analyzed (structure with rear wheels aligned and misaligned) and the results were compared. It was observed that the model with rear wheels misaligned showed maximum displacement lower than the model with rear wheels aligned. Although higher stress was found in the rear wheels misaligned, it was observed that average stresses for the misaligned wheels design were lower in most structural components analyzed. Based on FEM results, the coffee harvester machine with misaligned rear wheels was built and subjected to operational tests without showing any structural failure.

Keywords: Finite Elements Method; Stress Concentration; Static Simulation; Coffee Harvester

1. INTRODUCTION

Harvesting systems vary widely and the choice of one system over another will depend on labor availability, labor cost, topography and climatic conditions [1] . Harvest operation represents significant percentage of the total cost production and the mechanization provides quickly process, increasing the productivity and quality in harvest. The mechanization in harvesting coffee has been justified on several points, mainly in the economic aspect, arising efficient machines that require qualified labor to operate equipment in higher yields at harvest.

Numerical simulations in industries have been used since 70's. The use of numerical simulation can predict the behavior of the system under study in order to reduce risk in decision making, identifying problems before they occur, reducing costs in evaluating the technical and economic feasibility of a given project. It provides prediction of failure situations in machines during the product development phase which can accelerate manufacturing processes, reducing costs and timing.

Agricultural machinery can also be designed and simulated using specific platforms via commercial software. This kind of software works together with CAD (Computer Aided Design) that allows users to design and visualize the components in different views as well as modify and resize them. They also work with CAE (Computer Aided Engineering) platform which performs, among others, static simulations, allowing the best design in terms of structural integrity, generating cost savings in the product development phase.

The need for constant improvements in coffee harvesters, making them more efficient to meet products with quality, can affect physical structure and safety, which justify the use of numerical tools in the product development phase. Static simulations in the machine structure can predict failure situations in the field, by choosing the appropriated machine design.

This paper describes static analysis on the structure of a machine coffee harvester, considering two situations which are machines with rear wheels aligned and misaligned. These analyses were performed by using finite element method (FEM). FEM is a numerical method such as finite difference method but more general and powerful in its application to real-world problems. By using FEM for the machine structure in analysis, it was possible to generate values of stresses and displacements for both situations which allowed the choice of suitable design for structural components of a machine coffee harvester.

One trend that may help the development of complex machines to meet the needs of new practice in agriculture is the Virtual Prototyping (VP). VP is a CAD process that begins with the construction of models of digital products and performs simulations, addressing general issues of physical layout, operational concept, functional specifications and dynamic analysis in different operating environments [2] . CAD technology has spread throughout the world as a tool in the automotive, aeronautical and aerospace engineering, initiating CAE technologies that enable improved models by simulations of virtual machines performance. It is normally helped by FEM which has good agreement with experimental results in several applications [3] -[7] .

FEM is a technique used for computational analysis of stresses and vibrations and also for problems involving heat transfer, fluid mechanics and other applications [8] . FEM is based on the assumption that the displacement of each element may vary inside the body analyzed. To do a complete discretization, all loads applied to the structure must be condensed into a set of equivalent discrete nodal forces according to:

$$F = M\ddot{u} + C\dot{u} + Ku \tag{1}$$

where F is a nodal load vector, $M \cdot \ddot{u}$ is the inertial load vector, $C \cdot \dot{u}$ is the damping load and $K \cdot u$ is the elastic load vector.

Based on the loads acting in the machine as input data, is possible to predict the structure displacements helped by mathematical modelling such as FEM. Also helped by FEM and material properties information, is possible to predict the Von Mises (VM) stress, for ductile materials, which is associated to the distortion energy of components. A complete FEM bibliography review and practical applications are presented by [9] [10] .

The increase of computers processing power and the need for studies of complex machines are the main factors that contributed to the development of mathematical modelling by using FEM. Recently this method has been also used in agriculture engineering [11] -[14] . The present paper describes a novel application of FEM as an auxiliary tool for structural simulations of coffee harvester designs.

2. MATERIAL AND METHODS

The mechanized coffee harvest is based on vibration by canes placed in one or two vertical oscillating cylinders on the coffee harvester [15] . When the harvester is connected to a tractor which has a set of vibration operating one side plant per pass, the coffee detachment is done regardless of the alignment of coffee as result of the lateral adjustment of the set vibration. Through an inclined mechanism, it can operate even on slopes. The height adjustment enables the coffee detachment, working in trees of different sizes.

In self-propelled harvester, that is the subject matter of this paper, the involvement of the entire plant is done by the machine, vibration and continuous pathway coffee stem, seeking the detachment and subsequent gathering, cleaning and bagging. Has gantry structure, with sufficient height and width of plants. Driven by a diesel engine placed at the top of the structure resembles a tractor tall legs, with the components installed inside harvesting of its gantry. The machine operates in the line of plants, involving them fully.

The coffee harvester structure analyzed was designed and statically simulated by using the commercial software SolidWorks®. Figure 1 highlights the key parts of a self-propelled coffee harvester.

Figure 2 shows the CAD model isometric view for the original structure with each component identified.

In order to simulate different conditions of working in a coffee harvester machine, two situations had been used according to the following situations:

- Design A: full coffee reservoir and rear wheels aligned;
- Design B: full coffee reservoir and rear wheels misaligned.

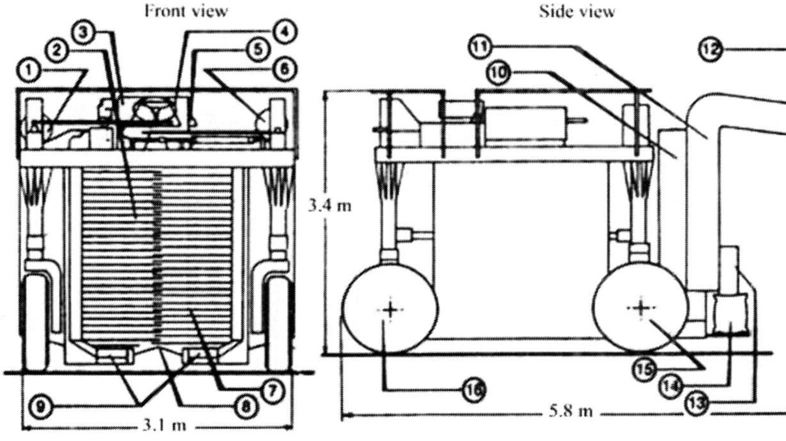

Figure 1. Self-propelled coffee harvester scheme. Legend: 1. Diesel tank; 2. vibrators; 3. diesel engine; 4. command post; 5. operational controls; 6. hydraulic system; 7. vibrators; 8. collectors; 9. carriers; 10.elevator; 11. cleaning system; 12. dirties output; 13. harvested coffee; 14.coffee bagging; 15. rear wheels; 16. front wheel.

The use of aligned and misaligned rear wheels in this study is justified by the focus on a new structural design of harvester coffee machines in order to provide a better efforts distribution through an unconventional rear wheels configuration using finite element resources. Figure 3 shows CAD models for coffee harvester structures complaining situations A and B (it is noted that some components were removed from the original design for a model simplifications during the simulations).

The material used was ASTM A36 steel for the structural parts and AISI 304 stainless steel for the fuel tank which the main properties are shown in the Table 1.

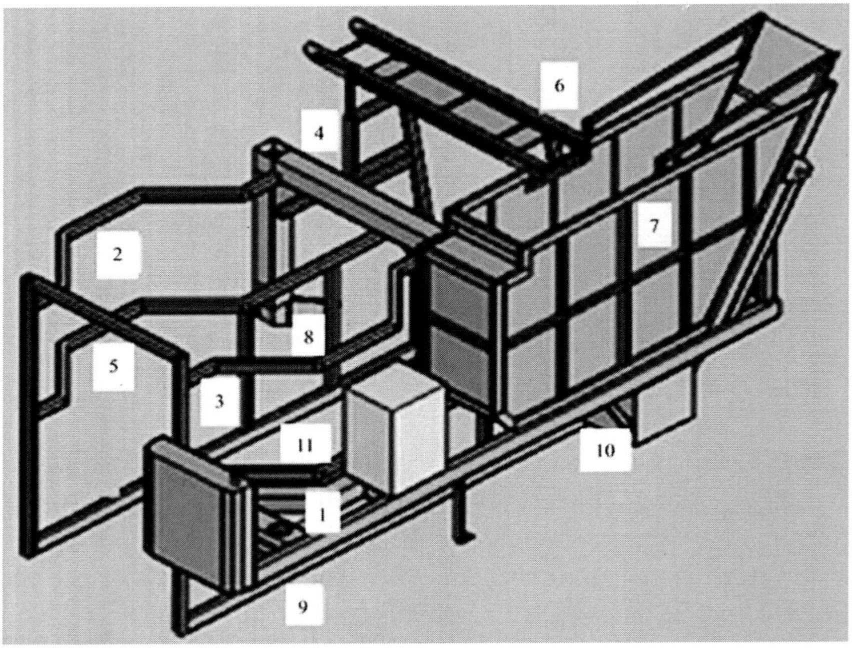

Figure 2. CAD model isometric view for the original structure. Legend: 1. Engine frame; 2. Body right side; 3. Body left side; 4. Main beam; 5. Body front end; 6. Body rear end; 7. Coffee reservoir; 8. Right rear wheel; 9. Guide wheel; 10 left rear wheel; 11. Fuel tank.

External loads acting in the machine structure were defined based on the own weight of each component and substances densities (Table 2). In addition to the substances weight, simulations were performed taking into account the weight of the components according to Table 3. Furthermore, clamping loads were defined according to attachment points of each component in analysis.

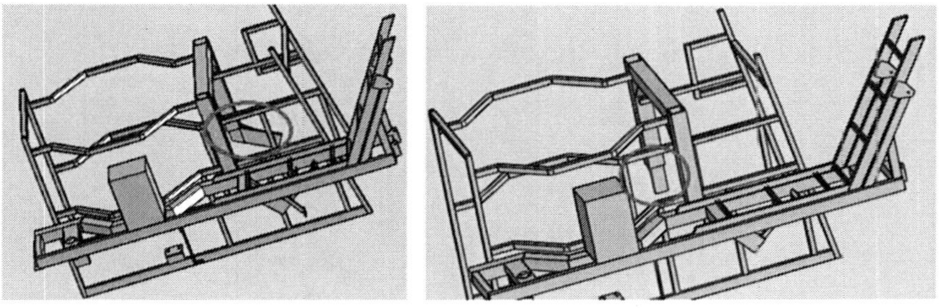

(a) (b)

Figure 3. Coffee harvester structure CAD models for (a) rear wheels aligned and (b) rear wheels misaligned.

Table 1. Material properties of structural parts and fuel tank.

Parts	Structural	Fuel tank
Material	ASTM A36	AISI 304
Density (kg/m³)	7850	8000
Poisson's ratio	0.26	0.29
Yield strength (N/m²)	2.5×10^8	2.1×10^8
Tensile strength (N/m²)	4.0×10^8	5.2×10^8
Elasticity modulus (N/m²)	2.0×10^{11}	1.9×10^{11}
Shear modulus (N/m²)	7.9×10^{10}	7.5×10^{10}

Table 2. Densities of substances.

Substance	Density (Kg/l)
Hydraulic oil	0.89
Fuel oil	0.85
Water	0.99
Coffee bean	0.70

Table 3. Loads applied in the machine structure.

Component	Load (N)
Full diesel tank	1440
Full radiator	500
Full coffee reservoir	15.600
Diesel engine	5000
Wheels components	2580
Panel and chair	1800
Battery	400
Oil reservoir	630
Pistons	1250
Gutters	4150
Full conveyor	4500
Vibrators and structure	9400

The mesh of machine structural parts was defined as triangular element type with ranging size between 1.2 mm and 47.0 mm which generated 35121 nodes. Figure 4 represents a portion of the generated mesh for the machine with rear wheels aligned.

3. RESULTS AND DISCUSSION

The last step was to simulate static analysis of the harvester machine in order to plot VM stress and displacement for the structure with rear wheels aligned and misaligned. Figure 5 shows the maximum VM stress for both situations.

Figure 6 shows the maximum displacement for each situation (displacement results for the machine with rear wheels aligned and rear wheels misaligned).

Figure 7 shows graphics that compares results of maximum VM stresses and average VM stresses.

Even though design B has presented highest maximum VM stresses (Figure 7(a)), it was observed lowest average VM stresses (Figure 7(b)) for the same design compared to design A, showing better stress distribution in the structure with rear wheels misaligned (design B) with 4,5% reduction of average VM stresses.

Figure 8 shows values of maximum displacements in design A compared to design B, which proved better performance of structure with rear wheels misaligned in terms of design.

In order to certify which situation would be the suitable design option, it was analyzed maximum VM stress compared to the average stress in each structural component for both situations (design A and design B) as shown in Figures 9 and 10.

Focusing on wheels zones stresses (Figures 9 and 10) it is noted, in both cases, that:

- The structure of left wheel presented higher values of maximum VM stress and the highest value of average VM stress compared to the other structural parts as result of the full coffee reservoir weight above the rear left wheel.
- The guide wheel showed low values of maximum VM stress and average stress even comparing with other structural parts, except for the body rear end.
- Average VM stress values for the misaligned wheels design were lower in 7 of 10 analyzed structural parts compared to the wheels aligned design.
- The structure of right wheel for design B (rear wheels misaligned) reduced by half the maximum VM stress when compared to design A (rear wheels aligned).

Based on that, it was certified that misaligned wheels design presented better performance than aligned wheels design in terms of stresses and displacements. The coffee harvester machine with misaligned wheels was built and started working without any structural fail as shown in Figure 11.

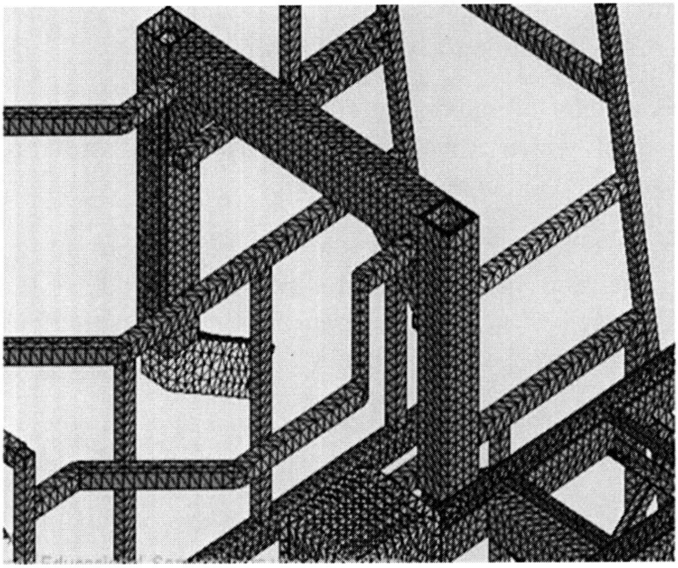

Figure 4. Portion of generated mesh for the harvester machine with rear wheels aligned.

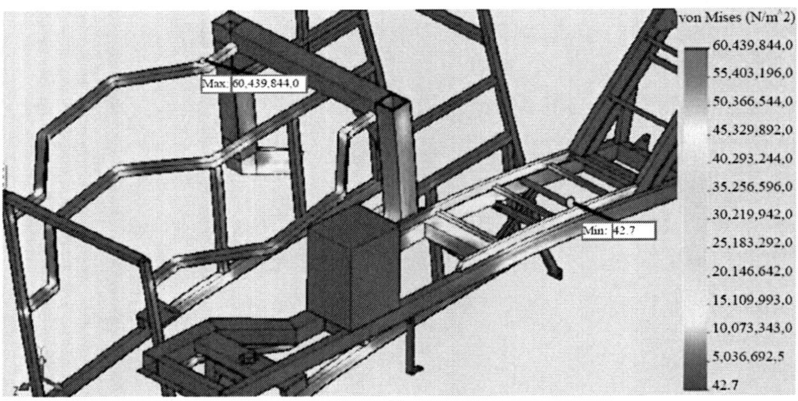

(a)

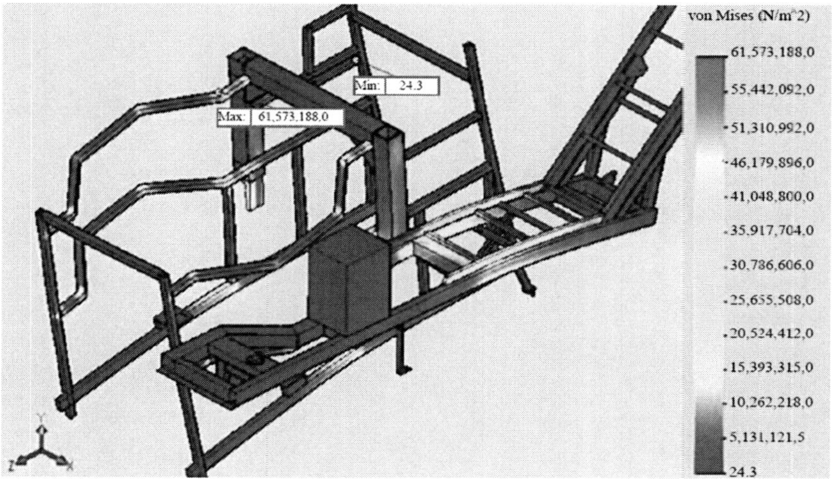

(b)

Figure 5. (a) VM stress results for the rear wheels al:gned; (b) VM stress results for the rear wheels misaligned.

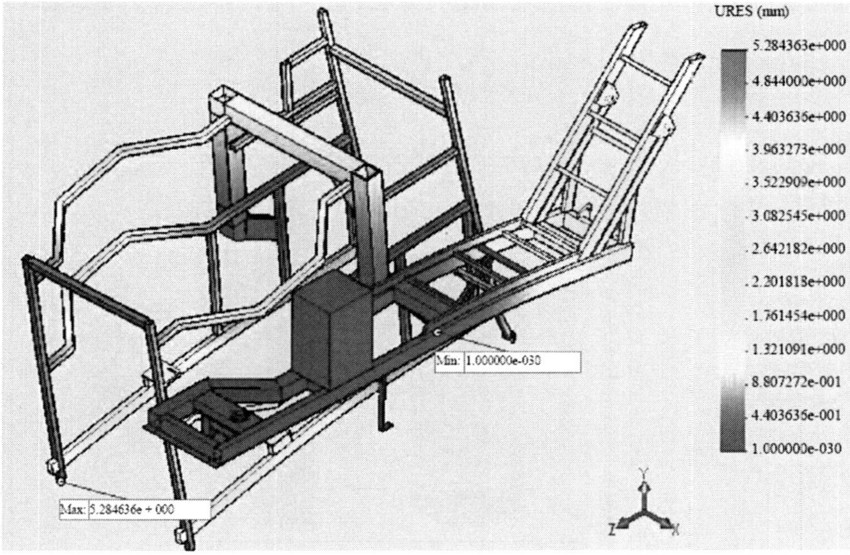

(a)

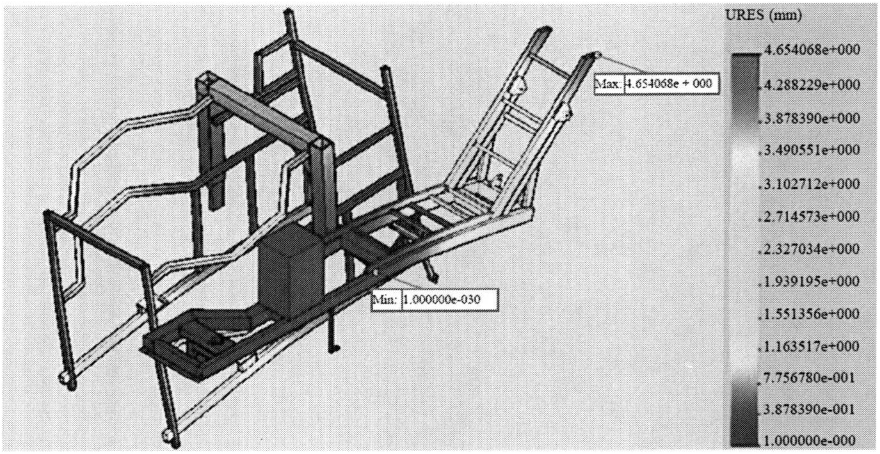

(b)

Figure 6. (a) Displacement results for the machine with rear wheels aligned; (b) Displacement results for the machine with rear wheels misaligned.

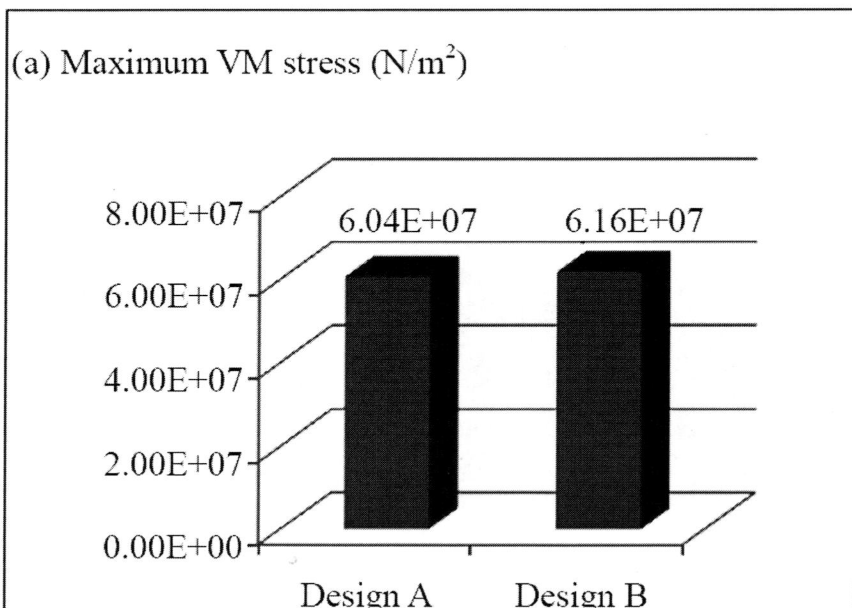

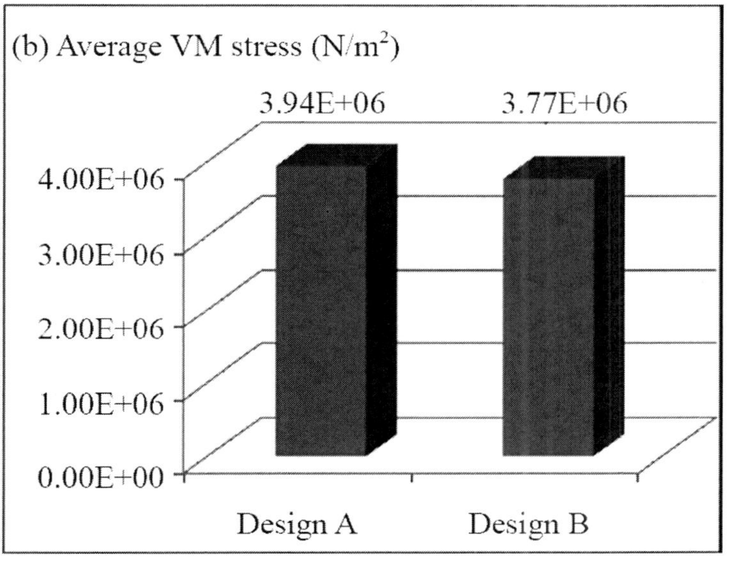

Figure 7. (a) Maximum VM Stresses; (b) Average VM Stresses.

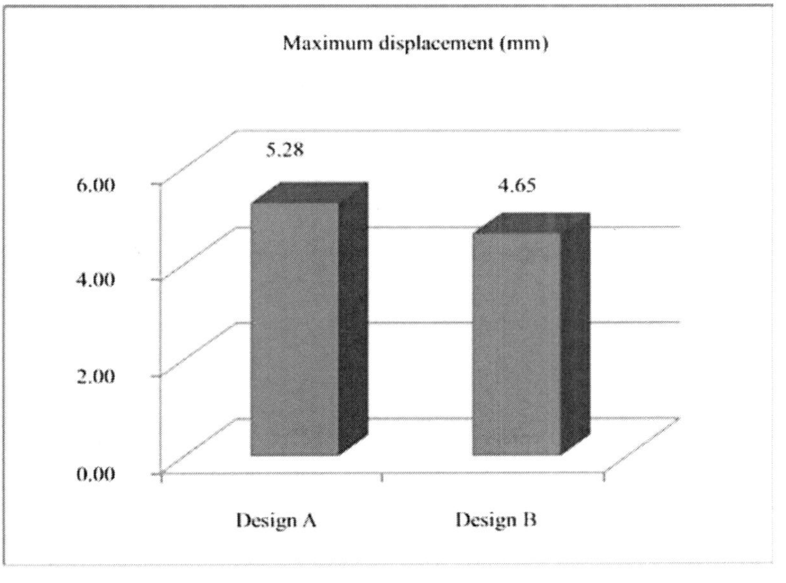

Figure 8. Maximum displacement for design A and design B.

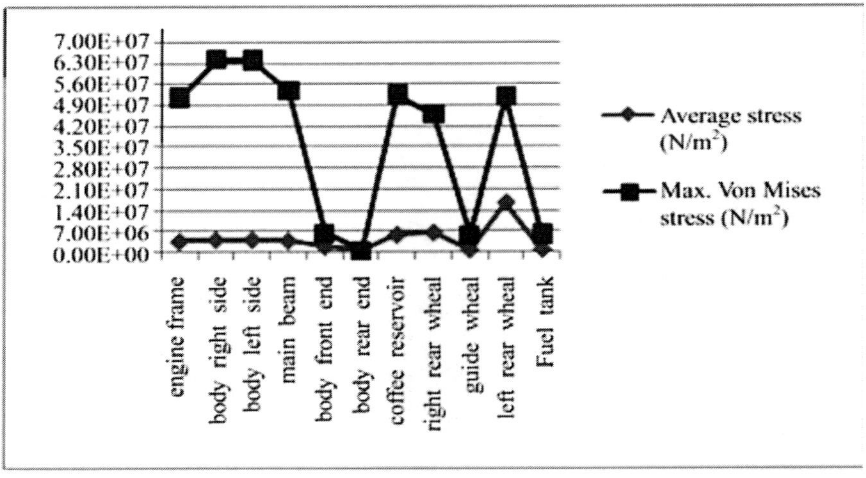

Figure 9. Maximum VM stress vs. average stress for the rear wheels aligned.

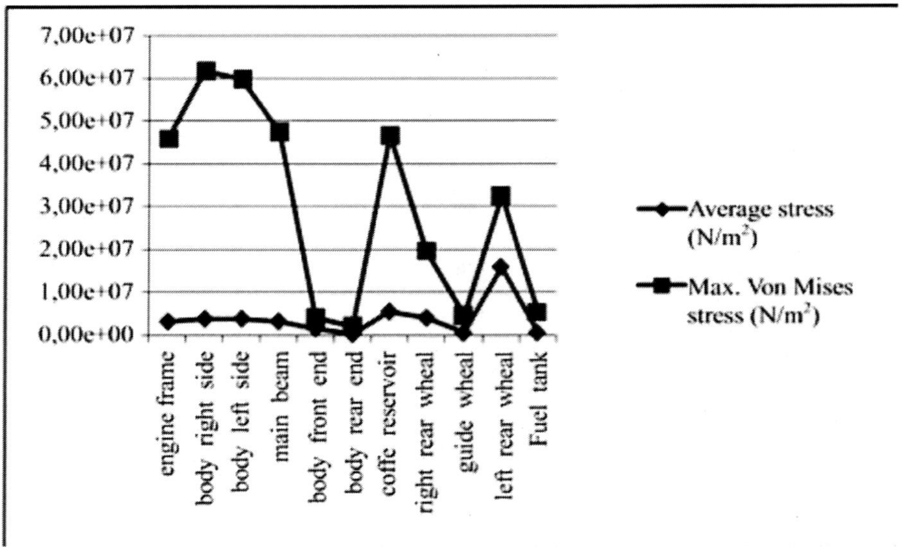

Figure 10. Maximum VM stress vs. average stress for the rear wheels misaligned.

Figure 11. Coffee harvester machine with misaligned rear wheels in operation.

4. CONCLUSIONS

In the presented paper, analysis of VM stress and displacement were supported by static finite element analysis. This paper had no intention to show any dynamic effects of a coffee harvester machine and this can be considered a limitation of the presented work. Also only structural parts were analyzed in this study as result of FEM timing consumed if finishing parts were added.

It was compared two different designs of a coffee harvester structures with rear wheels aligned and misaligned. FEM results showed maximum displacement in the structure with rear wheel misaligned (4.65 mm placed in the rear end structure) lower than the model with rear wheels aligned (5.28 mm placed in the front end structure) which represented a difference of 11.9% providing better machine structural equilibrium. Maximum VM stress results for the machine structure with rear wheels misaligned was 6.16×10^7 N/m^2 against 6.04×10^7 N/m^2 for the machine structure with rear wheels aligned, which represented a difference of 1.9%. Even though design with rear wheels misaligned presented highest maximum VM stresses, it was observed reduction of 4.5% in terms of average VM stresses.

Going in deep in each structural component for VM stress results, it was observed that the average stresses for the misaligned wheels design were lower in 7 of 10 analyzed parts compared to the wheels aligned design. In addition, the structure of right wheel for design with rear wheels misaligned reduced by half of the maximum VM stress when compared to design with rear wheels aligned. Based on FEM results, it was certified that the structure with rear wheels misaligned was the best design concept. It was confirmed in the field after a coffee harvester built with misaligned rear wheels with no fail during operation.

ACKNOWLEDGEMENTS

The authors would like to thanks the company VETOR for technical support and financial resources and CAPES for financial support for the project research.

REFERENCES

1. Thiyagarajan, R., Kathirvel, K. and Jayashree, G.C. (2013) Ergonomic Intervention in Sugarcane Harvesting Knives. African Journal of Agricultural Research, 8, 574-581.

2. Huang, T., Kong, C.W., Guo, H.L., Baldwin, A. and Li, H. (2007) A Virtual Prototyping System for Simulating Construction Processes. Automation in Construction, 16, 576-585.

3. Han, J.N., Cho, N.J., Chae, S.W. and Choi, Y. (2010) Stress Analysis of Ventilated Brake Discs Using the Finite Element Method. International Journal of Automotive Technology, 11, 133-138.

4. He, B., Wang, S. and Gao, F. (2010) Failure Analysis of an Automobile Damper Spring Tower. Engineering Failure Analysis, 17, 498-505.

5. Petracconi, C.L., Ferreira, S.E. and Palma, E.S. (2010) Fatigue Life Simulation of a Rear Tow Hook Assembly of a Passenger Car. Engineering Failure Analysis, 17, 455-463.

6. Wang, X. and Zhang, X. (2010) Simulation of Dynamic Cornering Fatigue Test of a Steel Passenger Car Wheel. International Journal of Fatigue, 32, 434-442.

7. Magalhaes, R.R., Fontes, C.H. and Melo, S.A.B.V. (2012) Failure Analysis and Design of a Front Bumper using Finite Element Method along with Durability and Rig Tests. International Journal of Vehicle Design, 60, 71-83.

8. Tang, C.Y., Chan, W. and Tsui, C.P. (2010) Finite Element Analysis of Contact Pressures between Seat Cushion and Human Buttock-Thigh Tissue. Engineering, 2, 720-726.

9. Zienkiewicz, O.C. (1970) The Finite Element Method: From Intuition to Generality. Applied Mechanics Reviews, 23, 249-256.

10. Knight, C.E. (1993) The Finite Element Method in Mechanical Design. Pws-Kent, Boston.

11. Magalhaes, A.C., Teixeira, M.M., Couto, S.M. and Resende, R.C. (2006) Modeling of Pneumatic Machine Collector of Coffee Fruits on Yard Using Finite Elements Analysis. Engenharia Agricola, 26, 483-492.

12. Celik, H.K., Rennie, A.E.W. and Akinci, I. (2011) Deformation Behaviour Simulation of on Apple under Drop Case by Finite Element Method. Journal of Food Engineering, 104, 293-298.

13. Nilnonta, W., Thepaa, S., Janjaib, S. and Kasayapananda, N. (2012) Finite Element Simulation for Coffee (Coffea Arabica) Drying. Food and Bioproducts Processing, 90, 341-350.

14. Li, Z., Li, P., Yang, H. and Liu, J. (2013) Internal Mechanical Damage Prediction in Tomato Compression Using Multiscale Finite Element. Journal of Food Engineering, 116, 639-647.

15. Ferraz, G., Silva, F.M., Alves, M.C., Bueno, R.L. and Costa, P. (2012) Geostatistical Analysis of Fruit Yield and Detachment Force in Coffee. Precision Agriculture, 13, 76-89.

11
CHAPTER

FINITE ELEMENT ANALYSIS OF VON-MISES STRESS DISTRIBUTION IN A SPHERICAL SHELL OF LIQUIFIED NATURAL GAS (LNG) PRESSURE VESSELS

Oludele Adeyefa, Oluleke Oluwole

Mechanical Engineering Department, University of Ibadan, Ibadan, Nigeria

ABSTRACT

This research work investigated the modeling of Von Mises stress in LNG Spherical Carbon Steel Storage tank using assumed displacement Finite Element analysis based on shallow shell triangular elements. Using equations of elasticity, constant thickness carbon steel spherical storage tanks were subjected to different loading conditions. This paper stresses the need for proper definition of shallow element using sector angles to obtain the shallowness. The shallow spherical triangular element has five degrees of freedom at each of its corner node, which are the essential external degrees of freedom. The assumed displacement fields of these shallow triangular elements satisfied the exact requirement of rigid body modes of motion. The FORTRAN 90 programming language was used for the programme coding to solve finite element equations resulting from the model while Von Mises stresses distribution within the spherical storage tank shell subjected to different internal pressures were determined. The results showed that the use of non-shallow elements due to improper sector angles resulted in unreliable results while real shallow elements produced results that tallied with ASME Section VIII Div 1, Part UG values.

Keywords: Von Mises Stress, FE Modeling, LNG, Spherical Storage Tank, ASME, ASCE

1. INTRODUCTION

Finite element modeling of pressure vessels response to stress analyses is a much inviting option than performing physical model analyses demonstrated by the amount of published work in the area. A large above-ground full containment LNG storage tank was modeled to check against local earthquake loading properties [1] while Korea Gas Corporation (KOGAS) developed the world's largest above-ground full containment LNG storage tank with a gross capacity of 200,000 m^3 [2]. The stability of cylindrical above-ground steel tanks under imposed support settlements and wind pressures have also been done [3]. In another study, the FEM code, MARC, was used to simulate the hydrobulging process of a singlecurvature polyhedron, including loading and offloading conditions [4]. Pascal Pourcel and others worked on the seismic post elastic behavior of an existing equipment with a volume of 1000 m^3 containing 85% of LPG [5]. In another study, the seismic response of a cylindrical steel liquid storage tank was examined by using a coupling method that combined the finite elements and the boundary elements [6]. However, in all cases of modeling, it is of utmost importance to select elements that would aptly describe the shape of the vessel for the element meshing and interpolation functions. Inapt defining of the element shape could be very disastrous in critical equipment modeling and design for applications. This paper looks at the critical area of using shallow shell element for finite element modeling of LNG spherical tanks.

2. FINITE ELEMENT THEORY

2.1. General

Assumed displacement method was employed in this research work to develop a shallow triangular spherical shell element without an in-plane rotation as a sixth degree of freedom.

Assumptions

- Uniform Pressure is assumed for the LNG storage tank.
- External wind load, seismic loads, tremors or earthquakes are not considered here.

2.2. Displacement Functions

The accuracy which may be obtained by the finite element method depends directly on the accuracy with which the deformation patterns are selected. The assumed deformation patterns should reproduce the distortions actually developed within the element. If deformation patterns are not properly chosen, the deformations will not necessarily converge to correct values when the mesh size is decreased. On the other hand, very good results may be obtained with a very coarse mesh if the element deformation patterns selected closely correspond to the actual patterns. Thus, the most critical factor in the entire finite element analysis is the proper selection of the element displacement field. To fulfill the conditions of the principle of minimum potential energy, the interpolation functions must be such that the displacements along the inter-element boundaries are compatible.

The assumed displacement relationships for the triangular shallow shell (Figure 1) are expressed in curvilinear coordinates. Polynomial displacement function is assumed; the polynomial being of the highest order that will permit evaluation of the coefficients. Since displacements u and v are known at three points, nodes 1, 2 and 3, the highest-order expressions which can be assumed for u and v are

$$u(x,y) = a_1 + a_2 x + a_3 y \tag{1}$$

$$v(x,y) = a_4 + a_5 x + a_6 y \tag{2}$$

The displacement w with its derivates, θ_x, θ_y has nine known values; hence, it may be assumed that

$$\begin{aligned} w(x,y) = {} & a_7 + a_8 x + a_9 y + a_{10} x^2 + a_{11} xy + a_{12} y^2 \\ & + a_{13} x^3 + a_{14} xy^2 + a_{15} y^3 \end{aligned} \tag{3}$$

from which it follows that

$$\theta_x(x,y) = \frac{\partial w}{\partial y} = a_9 + a_{11} x + 2a_{12} y + 2a_{14} xy + 3a_{15} y^2 \tag{4}$$

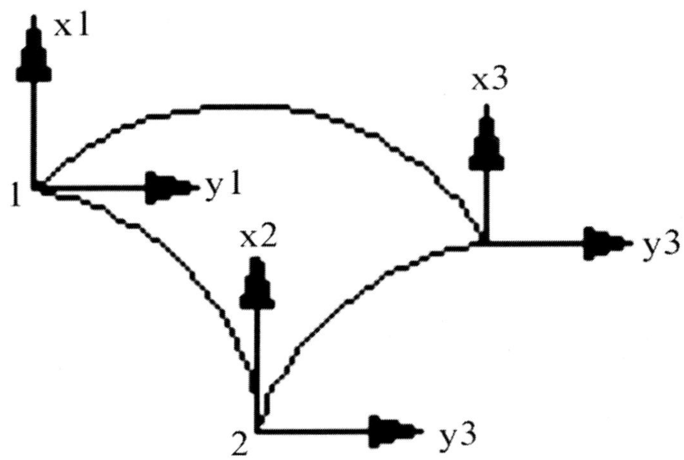

Figure 1. Shallow triangular element.

$$\theta_y(x,y) = -\frac{\partial w}{\partial x} = -a_8 - 2a_{10}x - a_{11}y - 3a_{13}x^2 + 2a_{14}y^2 \tag{5}$$

to determine constants as, known displacements at nodes are substituted and the equations become

$$[a] = \left[A^{-1}\right][\delta] \tag{6}$$

where [δ] is the nodal degrees of freedom, [A^{-1}] is inverse of transformation matrix and [a] is vector of independent constants.

2.3. Strain Displacement Relationship

The strain-displacement relationships for thin shells as given by Reissner [7] are simplified for the shallow shell and expressed as follows in curvilinear coordinates.

$$\varepsilon_x = \frac{\partial u}{\partial x} + \frac{w}{r}, \varepsilon_y = \frac{\partial v}{\partial y} + \frac{w}{r}, \varepsilon_{xy} = \frac{\partial u}{\partial y} + \frac{\partial v}{\partial x},$$

$$k_x = -\frac{\partial^2 w}{\partial x^2}, k_x = -\frac{\partial^2 w}{\partial y^2} \tag{7}$$

The strain equations can be written in matrix form after necessary substitutions of u, v and w into the above strain equations.

2.4. Stress in a Triangular Element

Stress varies from point to point along the shell profile and also through the thickness of the shell. It is thus in reality an unknown function of two variables, which leads us to the equations below:

$$\sigma_b = \frac{6M}{t^2}, \quad \sigma_m = \frac{N}{t} \tag{8}$$

where: M is the moment per unit length, M and σ_b is the bending stress at the surface.

N is to be force per unit length and σ_m which is membrane stress.

2.5. Strain Energy

The strain energy of an isotropic linear shell is given by Langhaar [8] as;

$$U = \iint_A \int_{-\frac{t}{2}}^{\frac{t}{2}} \frac{E}{2(1-v^2)}\left[\varepsilon_x^2 + \varepsilon_y^2 + 2v\varepsilon_x\varepsilon_y + \frac{1}{2}(1-v)\gamma_{xy}^2\right] \mathrm{d}\varsigma \mathrm{d}x \mathrm{d}y \tag{9}$$

where, t = thickness of the shell, v = Poisson's ratio and E = Modulus of elasticity.

After substitution for strains in the above expression and integration with respect to ς, the strain energy can be separated into the membrane energy U_m and the bending energy U_b.

$$U = U_m + U_b \tag{10}$$

$$U_m = \frac{Et}{2(1-v^2)} \iint_A \left[e_x^2 + e_y^2 + 2ve_x e_y + \frac{1}{2}(1-v) e_{xy}^2 \right] dxdy \tag{11}$$

$$U_b = \frac{Et^3}{24(1-v^2)} \iint_A \left[k_x^2 + k_y^2 + 2vk_x k_y + \frac{1}{2}(1-v) k_{xy}^2 \right] dxdy \tag{12}$$

The potential energy is then written as:

$$\Phi = U - W$$

where W represents the work done by the external load on the system. In the finite element method, the potential energy of a shell is expressed as:

$$\Phi = \sum_{k=1}^{n} \phi_k \tag{13}$$

where ϕ_k is the potential energy of the kth element.

2.6. Stiffness Matrix

By writing strain energy equations in terms of displacements, element stiffness matrix can be determined in the usual manner,

$$k_m = t\left[A^{-1}\right]^T \iint_A B_m^T D_m B \mathrm{d}x\mathrm{d}y \left[A^{-1}\right] \quad (14)$$

$$k_b = t\left[A^{-1}\right]^T \iint_A B_b^T D_b B \mathrm{d}x\mathrm{d}y \left[A^{-1}\right] \quad (15)$$

k_m and k_b are element stiffness matrices due to membrane and bending stresses respectively.

D_m and D_b are elasticity matrices for membrane and bending stresses respectively.

B_m and B_b are strain matrices for membrane and bending stresses respectively.

Therefore, element total stiffness matrix is

$$k = k_b + k_m \quad (16)$$

Element stiffness matrix is then combined to give system stiffness matrix.

2.7. Consistent Load Vector

The simplest method to establish an equivalent set of nodal forces is the lumping process. An alternative and more accurate approach for dealing with distributed loads is the use of a consistent load vector which is derived by equating the work done by the distributed load through the displacement of the element to the work done by the nodal generalized loads through the nodal displacements. If a triangular shell element is acted upon by a distributed load q per unit area in the direction of w, the work done by this load is derived as follows:

$$P_1 = \int_A qw\mathrm{d}x\mathrm{d}y \tag{17}$$

If w is taken to be represented by:

$$\{w\} = \left[C^T\right]\{a\} = \left[C^T\right]\left[A^{-1}\right]\{\delta\} \tag{18}$$

where [C^T] for the present element is given by

$$C^T = \left[0\ 0\ 0\ 0\ 0\ 0\ 1\ x\ y\ x^2\ xy\ y^2\ x^3\ xy^2\ y^3\right] \tag{19}$$

The work done by the consistent nodal generalized force through the nodal displacements

{δ} is given by:

$$P_2 = \left\{F^T\right\}\{\delta\} \tag{20}$$

Hence, from Equations (18)-(20), the nodal forces are obtained

$$F = \left[A^{-1}\right]\int\left[C^T\right][q]\mathrm{d}x\mathrm{d}y \tag{21}$$

Equation (21) gives the nodal forces for a single element; and the nodal forces for the whole structure is obtained by assembling the elements' nodal forces.

2.8. Boundary Conditions

In order to reduce computing time, symmetrical nature of the storage tank was considered. Shown below is the typical shallow spherical triangular shell mesh for finite element modeling of LNG spherical storage tanks. This mesh in the Figure 2 below has six shallow spherical triangular shell elements with eight nodes.

Each node has five degree of freedoms; therefore the mesh in Figure 2 has forty degrees of freedom. In considering known displacements, all displacements at given node were given zero values with the exception of radial displacements, w.

2.9. Sector Angles

It can be taken that there are many lines of symmetry in the spherical tank with uniform internal/external pressure loading. For the purpose of the finite element analysis, it is better to take advantage of this symmetry by considering portions of the spherical tank instead of the whole tank. Thus, if we used a quarter of the spherical tank to analyze the entire system, to know the size of this quarter of the spherical tank in terms of angle size, we take 1/4 of 360°—which is equal to 90°. The calculation of angle size of symmetrical portion of the spherical tank used for finite element analysis with respect to the whole tank is what we termed "sector angle".

Figure 2. Typical spherical shell mesh.

3. PROBLEM CONSIDERED

The direct stresses, N_x, N_y and the bending moment, M_x and M_y were computed along x and y axes respectively. Maximum Von Mises stresses are determined. The spherecal storage tank considered has the following properties:

Diameter = 40 in

Thickness = 1 in

Young Modulus of Elasticity = $30*10^6$

Psi Poisson Ratio = 0.3

Shell Material = A516M Grade 70

Shell Material Minimum Yield Stress = $38*10^3$ psi

Results and Discussions

The maximum Von-Mises stresses variations with internal pressures are given in Figure 3. Here, the shallow shell element used in modeling was well described at a sector angle of 90 degrees. Thus, the Finite element values showed close tally with ASME Section VIII Div 1, Part UG values. Equivalent maximum Von mises stresses were well within the limits of shell material minimum yield stress for internal pressure between 500 psi and 3800 psi. For internal pressure above 3800 Psi, the equivalent maximum Von mises yield stresses exceeded the shell material minimum yield stress.

To prevent failure of LNG storage tank at pressures above 3800 psi, one of the following has to be done.

- The shell thickness has to be increased accordingly so that equivalent maximum Von mises stress will be lower than the shell material minimum specified yield stress.

- Select shell material with higher minimum yield value so that the equivalent maximum Von mises stress calculated will be less than the shell material minimum specified yield stress.
- Decrease the working internal pressure of the spherical storage tank so that the equivalent maximum Von mises stress will fall below the shell material minimum specified yield stress.

Figure 4 shows the result when an element does not describe well the structure being modeled. Using a sector angle of 25 degrees, the shell element was not shallow but curved. This showed a deviation from ASME, s value by about 20 percent. The figure also shows that increasing the number of elements (i.e. making the element smaller) will not solve the problem of a bad choice of element configuration. The values kept diverging away from ASME standard values.

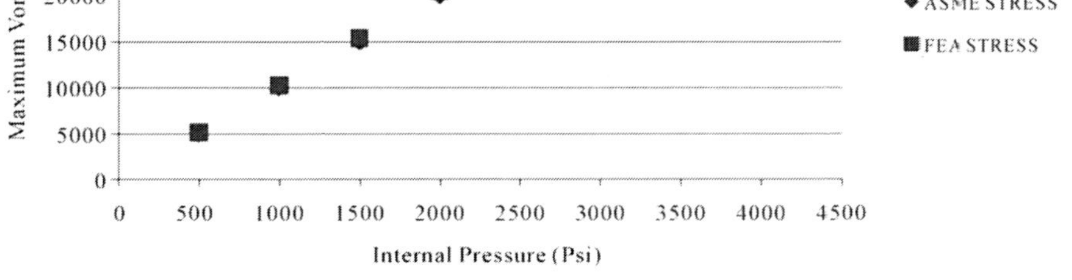

Figure 3. Graph of internal pressure (Psi) against maximum Von-Mises stress (Psi) for spherical vessel dia. of 40 inches and sector angle of 90 degrees producing a shallow shell element.

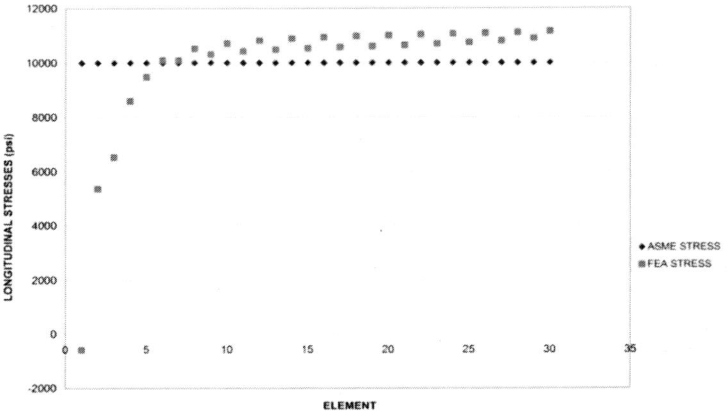

Figure 4. Plot of number of mesh elements against longitudinal stresses (Psi) using a sector angle of 25 for a 40 in Dia. Vessel and internal pressure of 1000 Psi (sector angle here produced a curved element and not a shallow shell element).

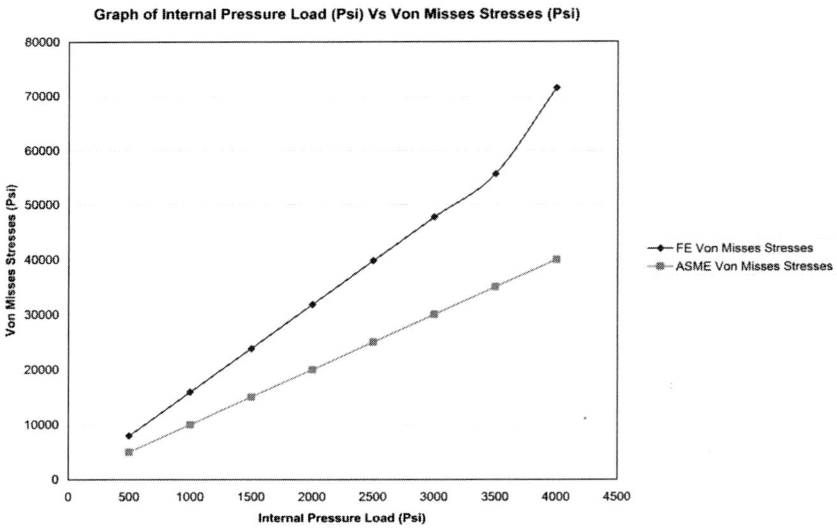

Figure 5. Internal pressure loading against Von Mises stresses using a sector angle of 25 degrees and hence curved shell element.

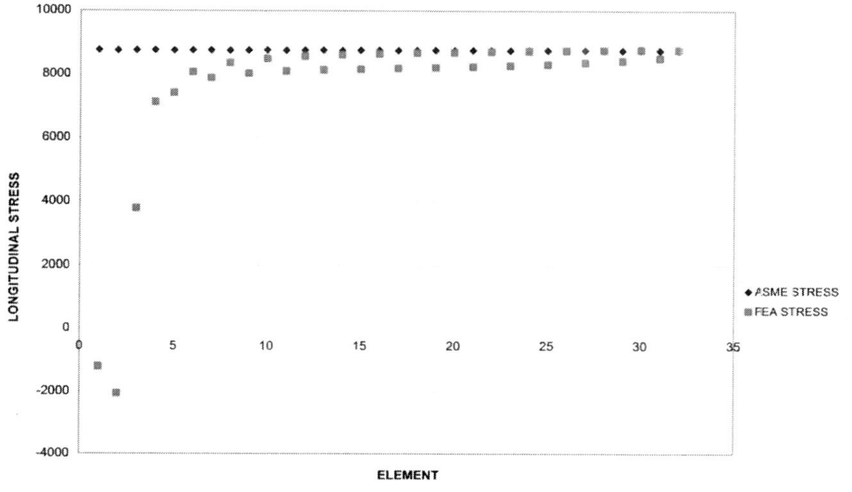

Figure 6. Plot of number of elements against longitudinal stresses for a 35 in Dia vessel modeled with a 30 degrees sector angle shell element showing negative deviation from ASME values.

The onus is thus on the design engineer to take responsibility for proper and painstaking modeling for design whichever modeling method he/she chooses.

4. CONCLUSIONS

The usefulness of finite element modeling of spherical steel shells for LNG pressure storage vessels using shallow shell elements has been indicated in this work, the usefulness of ASME standards as guides cannot be over emphasized. Strict adherence to adaptation of physical parameters to modeling parameters is indicated as seen in the use of shallow shells which aptly describe the tank structure as opposed to curved elements. A wrong choice of mesh element configuration will not make results converge to correct results no matter the discretization size. Design Engineers must carry out their design and modeling base on storage tank capacities and maximum loads that may act on the spherical storage tank during its design life.

5. REFERENCES

1. J. H. Kim, et al., "Preliminary Earthquake Response Analysis of 200,000 Kilolitres Large Capacity aboveGround LNG Storage Tank for the Basic Design Section," Institute of Construction Technology, DAEWOO E&C Co., Ltd., 2005.

2. J.-H. Kim, H.-S. Seo, K.-W. Lee and I.-S. Yoon, "Development of the World's Largest above-Ground Full Containment LNG Storage Tank," 23rd World Gas Conference, Amsterdam, 6 June 2006.

3. E. M. Sosa, "Computational Buckling Analysis of Cylindrical Thin-Walled above-Ground Tanks," Ph.D. Thesis, The University of Puerto Rico Mayaguez Campus, Mayaguez, June 2005.

4. J. Dong, et al., "Numerical Calculation and Analysis of Single—Curvature Polyhedron Hydro-Bulging Process for Manufacturing Spherical Vessels," Institute of Nuclear Energy Technology, Tsinghua University, Beijing 2005.

5. P. Pourcel, et al., "A Seismic Post Elastic Behaviour of Spherical Tanks," TECHNIP France and DYNALIS France, 1999, pp. 1-14.

6. Y. Dong and D. Redekop, "Structural and Vibrational Analysis of Liquid Storage Tanks," Transactions, SMiRT 19, Department of Mechanical Engineering, University of Ottawa, Ottawa, 2007.

7. E. Reissner, "On Some Problems in Shell Theory," Proceedings, 1st Symposium on Naval Structural Mechanics, Stanford University, Pergaman Press Inc., New York, 1960.

8. H. L. Langhaar, "Energy Methods in Applied Mechanics," Wiley & Sons, New York, 1962.